IMAGES
of America

NEWPORT BEACH FIRE DEPARTMENT

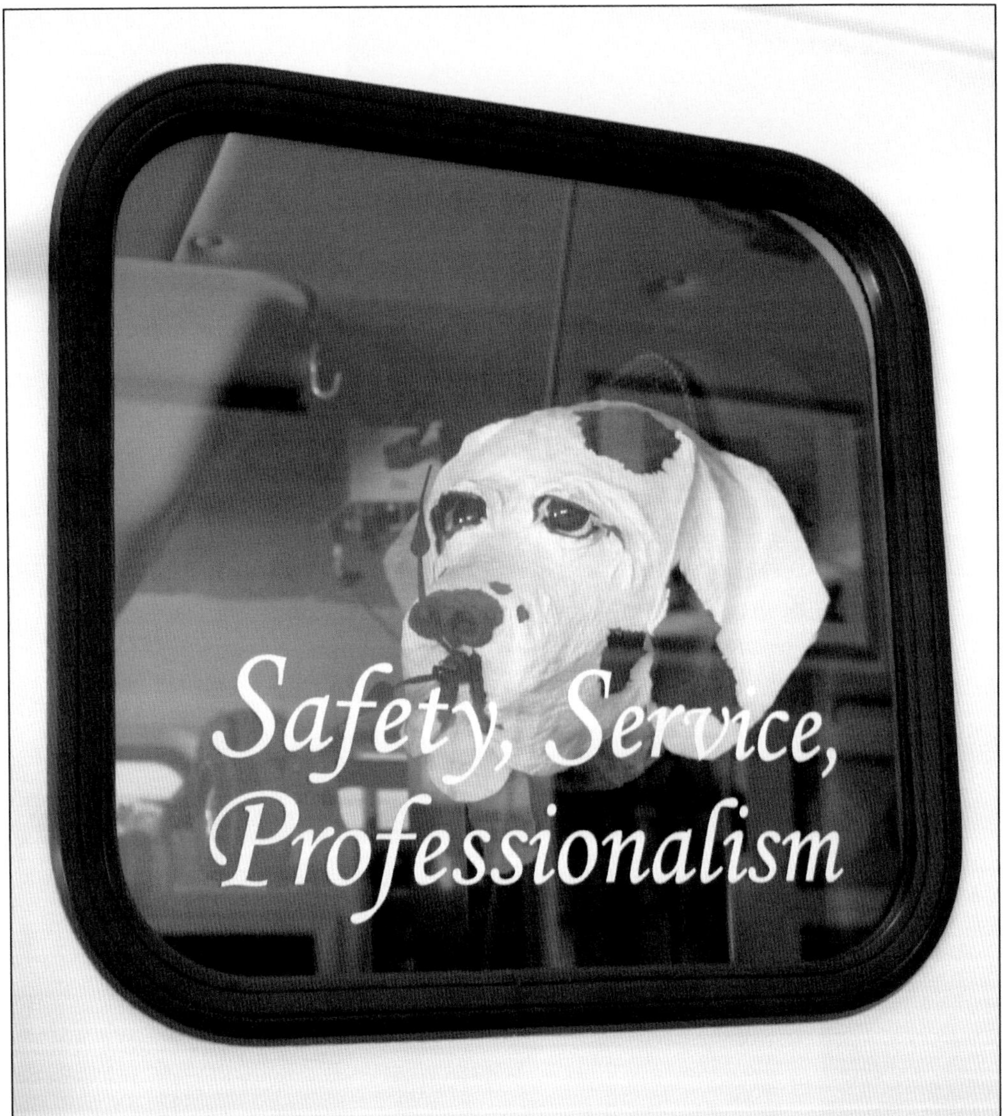

A papier-mache Dalmatian peeks out from behind the driver's seat of a Newport Beach fire engine. *Safety, Service, Professionalism* has been a guiding principle and motto of the Newport Beach Fire Department for 100 years. (Author's collection.)

ON THE COVER: Perce Pellett and Orlando Briscoe pose with Newport Beach's 1927 American LaFrance pumper in front of Fire Station One in 1937. (Courtesy of NBFD archives.)

IMAGES
of America

NEWPORT BEACH FIRE DEPARTMENT

Michael J. Novak

ARCADIA
PUBLISHING

ISBN 978-0-7385-5593-5

Published by Arcadia Publishing
Charleston SC, Chicago IL, Portsmouth NH, San Francisco CA

Printed in the United States of America

Library of Congress Catalog Card Number: 2007933021

For all general information contact Arcadia Publishing at:
Telephone 843-853-2070
Fax 843-853-0044
E-mail sales@arcadiapublishing.com
For customer service and orders:
Toll-Free 1-888-313-2665

Visit us on the Internet at www.arcadiapublishing.com

This volume is dedicated to the men and women of the Newport Beach Fire Department, past, present, and future.

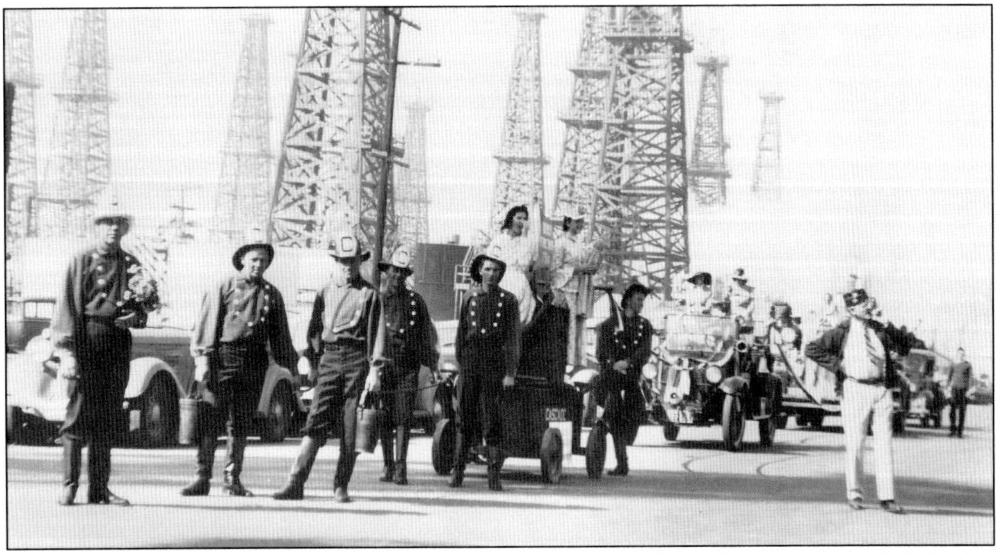

Assistant chief Perce Pellett leads members of the Newport Beach Fire Department and the Lady's Auxiliary in a parade along Pacific Coast Highway in 1938. (Courtesy of NBFD archives.)

CONTENTS

Acknowledgments

The work of three Newport Beach Fire Department firefighters paved the way for this volume. Chief Frank Crocker was a meticulous record keeper and compiler of early department history and an intrepid photographer. Chief Jan Briscoe, also an amateur photographer, saved everything from the early days in a priceless scrapbook. Capt. William "Dutch" Van Horn tracked down a photograph of virtually every member who wore a NBFD uniform and created a series of notebooks rich in department lore. Generations of history would have been lost to time if it were not for the efforts of these three individuals.

The NBFD and I owe a special thanks to Richard Koehler for opening up his collection of photographs for use in this volume. "Scoop" Koehler has documented 50 years of NBFD history as a professional news photographer for several prominent Southern California newspapers. A lifelong resident of Newport Beach, Scoop, camera in hand, started chasing Newport fire engines to calls on a single-speed bicycle as a teenager in the late 1950s. He is still at it.

I would like to thank fire chief Steve Lewis for his encouragement and support. Chief Lewis, Capt. Jerry Strom, and paramedic Kelly Gorman were nice enough to look this manuscript over with a critical eye. Thanks also to my sister Pat Novak for that final editorial touch. Jeff Remmele, Bob Kelly, Brent Jacobsen, Ron Gutierrez, Don Crocker, the Ray Shannon family, Tim Riley, Jerry Strom, Jen Schulz, and the Long Beach Fire Department Museum deserve special thanks for their contributions. Finally, I'd like to thank my wife, Robin, for taking care of the present while I lived in the past. Thank you one and all.

INTRODUCTION

The city of Newport Beach, California, sits along seven miles of Orange County's finest coastline, often referred to as the Gold Coast. Newport Beach is home to miles of white sand beaches, the finest small boat harbor on the West Coast, and some of the most expensive real estate in the country.

During the 1870s, Newport Bay was a difficult waterway to navigate. Intrepid pioneers established a boat landing beneath the Castaways Bluff, where supplies and building materials could be unloaded and transferred to wagons. The McFadden brothers took over the landing in 1875. Immediately, steamers began bringing in the lumber eagerly sought by builders in the growing city of Santa Ana a few miles inland.

In the late 1880s, the McFaddens, frustrated with the unreliability of the inland waterway and lawsuits involving landowner James Irvine, constructed a pier on the Balboa Peninsula. A rail line was extended to the pier. Ships could now unload their goods in a few hours and be on their way. A small village, called Newport, developed around the wharf. A few miles to the south, Balboa Village began to grow around the well-known Balboa Pavilion, built in 1905. In 1906, these two villages were incorporated into the city of Newport Beach.

The Pacific Electric Red Cars made their way as far as Balboa in 1906, bringing vacationers and investors to the peninsula beaches. Across the bay from the pavilion, a real estate developer by the name of William Collins was building up a sandbar called Balboa Island and selling lots ideal for summer cabins.

In August 1910, fire damaged the meeting room of the Newport Beach City Trustees. Suddenly, the threat of fire was foremost in the minds of both local politicians and residents. On April 17, 1911, Ordinance No. 65, calling for the creation of a volunteer fire department, was passed by the trustees. A volunteer fire chief was appointed. Three hose reels were purchased, along with fire hose. A fire hall was soon built a block away from the pavilion. Many residents and local businessmen quickly joined the ranks of the new department. These volunteers would prove to be a lively bunch. Regular meetings were held, along with dinners and dances at local establishments such as the Balboa Hotel. Bake sales were held to raise money for needed equipment.

In December 1915, Chief Herbert Castle sent a letter to the Long Beach Fire Department inquiring about the availability of two of their used motorized hose wagons. These vehicles, each built on a Rambler chassis, were purchased and delivered to Newport and Balboa in 1916. The Newport Beach Fire Department now had its own motorized fire apparatus.

The best intentions of Newport Beach volunteers were often foiled by the lack of reliable equipment and a primitive city infrastructure. The following description of a typical fire illustrates the difficulties they encountered. In April 1918, several houses caught fire near Balboa Village, including the home of William Collins. An attempt to summon firefighters with the fire bell was thwarted when the bell's rope broke. Attempts to call Newport Village for help failed when it was discovered that the only telephone line in the area had been damaged by the fire. Balboa volunteers laid out a hose line but found the water pressure to be inadequate. Here, at least, luck was on their side. A call had been put in to the pump house manager just before the telephones went dead. "Pump for all you're worth," was the message. An eager resident jumped into his car and sped up the beach to sound the alarm in Newport, leaving a cloud of dust in his wake. Engine No. 1 arrived from Newport Village just in time to prevent an entire block of cottages from going up in flames. In the end, the zeal, dedication, and determination of the volunteers overcame all challenges and carried the day.

A pivotal year for the firefighters was 1920. A. W. Jackson was appointed volunteer fire chief. Chief Jackson would remain on as fire chief until 1927, giving the department much needed stability at the top. Jackson was committed to upgrading equipment and increasing the training of volunteers. He worked tirelessly at getting city trustees and residents interested in approving bonds that would provide funds for a new fire engine and fire station.

Nevertheless, bond measures continued to fail. Newport Engine No. 1 was destroyed in a fire in 1921. Several other destructive fires painfully highlighted the inadequacy of fire protection in the growing community. Ultimately, it was a mandated survey of the city's firefighting capabilities that spurred the city into action. The abysmal grade Newport received jeopardized favorable insurance rates. Fearing a backlash of criticism from city residents, the trustees scrambled for a solution.

In March 1927, Ordinance No. 315, authorizing the creation of a paid fire department, was passed and adopted by the City of Newport Beach. An American LaFrance Type 75 fire engine was purchased. A new fire station was built and opened in June. A paid fire chief and a fire apparatus driver were hired.

These two individuals, Frank Crocker and Clinton Dickey, became the cornerstones of the present-day Newport Beach Fire Department. The NBFD, now numbering over 150 men and women, has grown into one of the finest fire departments on the West Coast. Over the last 80 years, generations of firefighters wearing the NBFD badge have built the department into the professional organization it is today. Hard work, innovative thinking with an eye toward the future, and dedication to the residents of the city of Newport Beach continue to build upon a proud tradition. Eight fire stations are strategically located throughout the city. Three paramedic vans, two ladder trucks, and eight engines cover the city. Firefighters utilize the latest in sophisticated equipment to provide advanced emergency medical care. Fires, when they occur, are quickly extinguished with effective teamwork and reliance on sound training. A Fire Prevention Bureau conducts safety inspections and sees that codes designed to protect building occupants are rigidly enforced. Disaster preparedness programs are developed and Community Emergency Response Teams are trained. Along the shoreline and on the bay, Newport Beach Lifeguards, under the administration of the fire department, keep the beaches safe for visitors and residents.

One

The Volunteer Years

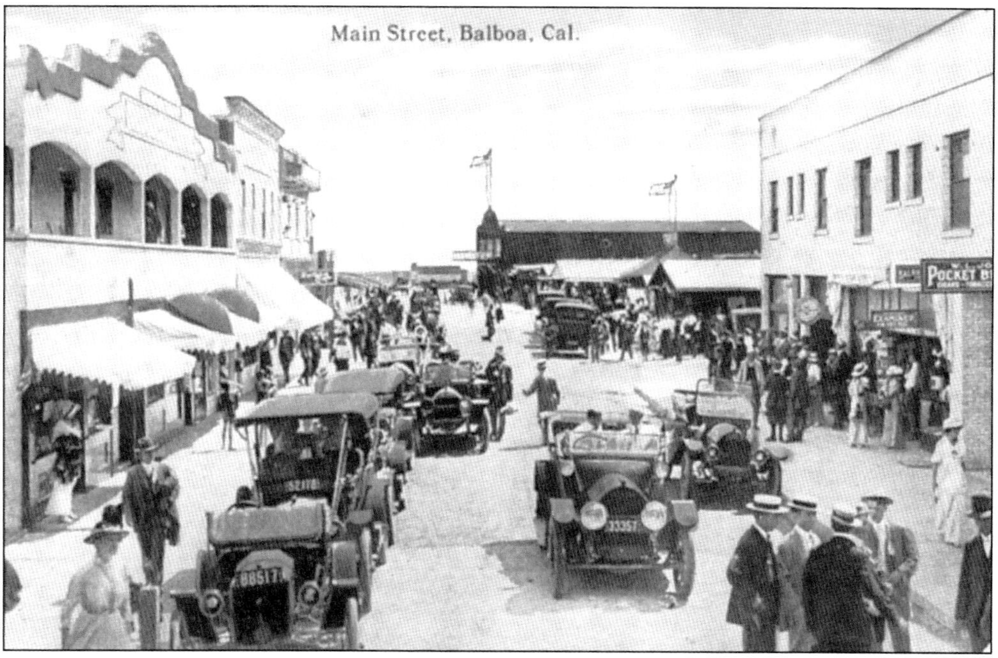

The city of Newport Beach was incorporated in 1906. In 1910, the city was made up of several small villages separated by a few miles of undeveloped oceanfront real estate. Two communities were emerging on the Balboa Peninsula, a five-mile-long stretch of sand whose wide beaches separated the Pacific Ocean from the tranquil waters of Newport Bay. Balboa Village, seen here, developed around the well-known Balboa Pavilion, built in 1905. Two miles up the beach, Newport Village was a thriving commercial center adjacent to the Newport Pier and a small fishing fleet. The famous Pacific Electric Red Cars brought visitors to both villages. (Author's collection.)

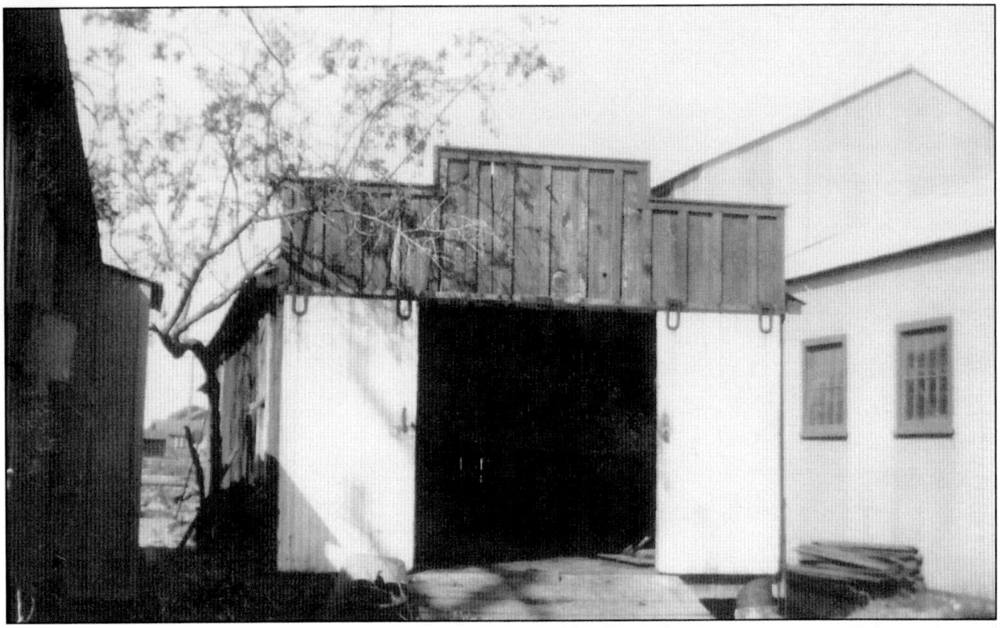

In August 1910, the city council chambers were ravaged by fire. The city trustees, rudely awakened to this threat, quickly sought to establish organized fire protection for the fledgling villages of Balboa and Newport. A vacant lot was purchased near the Newport Pier. Newport's first firehouse, a small shack, was erected on that lot. (Courtesy of NBFD archives.)

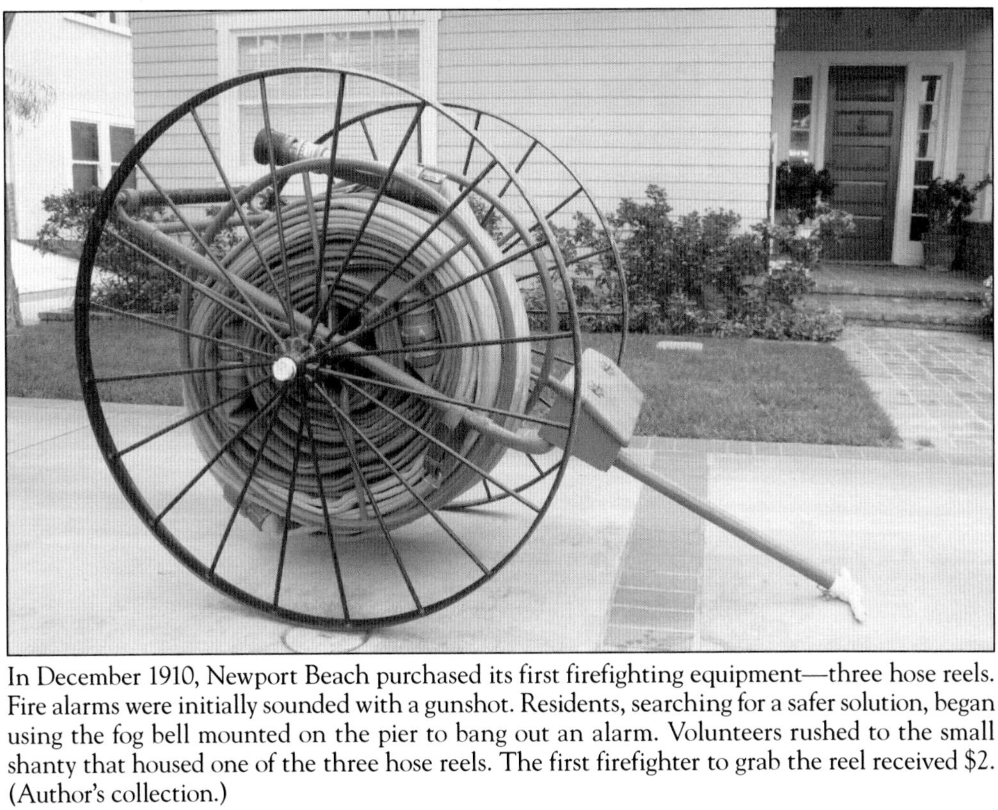

In December 1910, Newport Beach purchased its first firefighting equipment—three hose reels. Fire alarms were initially sounded with a gunshot. Residents, searching for a safer solution, began using the fog bell mounted on the pier to bang out an alarm. Volunteers rushed to the small shanty that housed one of the three hose reels. The first firefighter to grab the reel received $2. (Author's collection.)

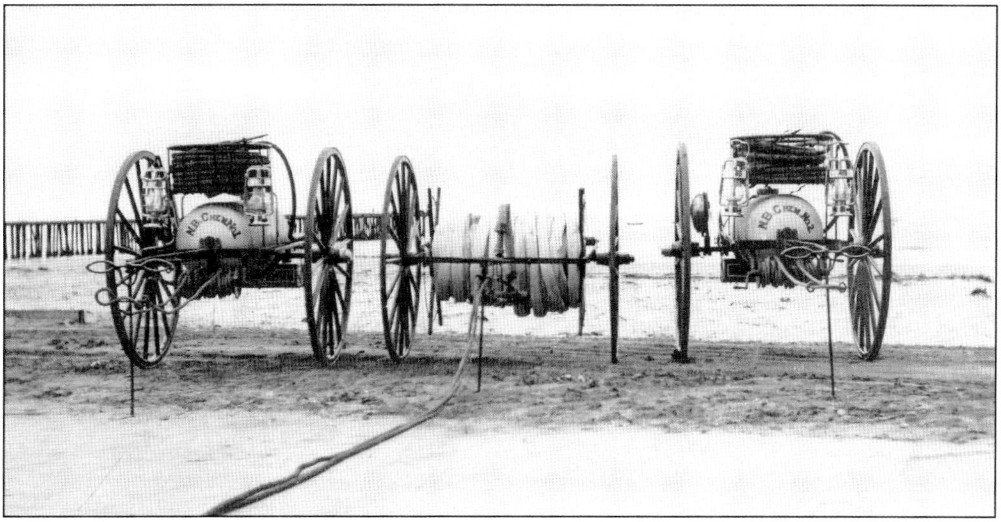

Two additional hose reels were purchased in January 1911. These reels carried 60-gallon water tanks and cost $760. On April 17, 1911, Ordinance No. 65, creating a volunteer fire department, was officially passed by the trustees. (Courtesy of NBFD archives.)

This photograph is purported to be of Newport's first volunteer fire chief, W. A. Cornelius, appointed in January 1911. Cornelius was editor of Newport's first newspaper, the *Newport News*. Chief Cornelius and his 15-member department received no pay. Cornelius resigned at the end of the year. He was replaced by Mannie Simberg, a local photographer who owned a business in Newport Village. City trustees immediately sought to make participation in the fire department more appealing by voting to pay the fire chief $10 a month and each member $2.50 per fire call. (Courtesy of NBFD archives.)

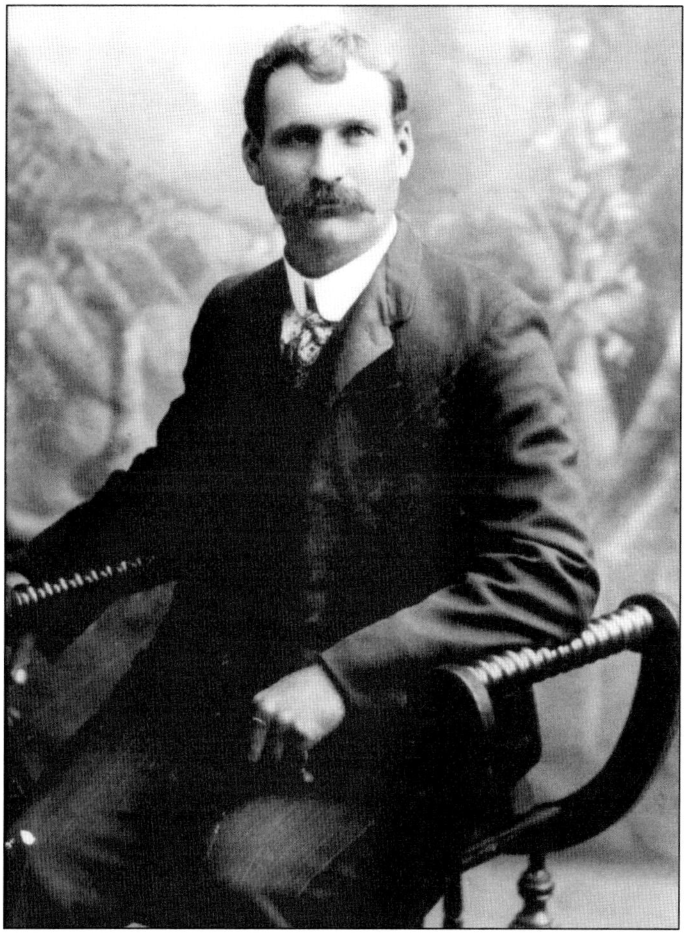

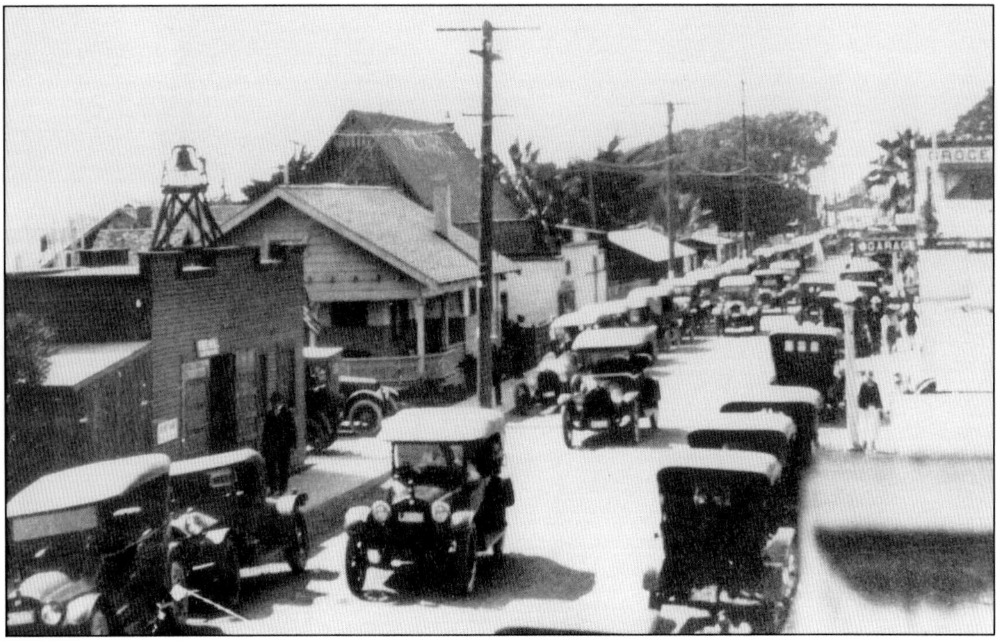

Around 1912, a new fire hall was constructed on East Bay Street in Balboa Village. The nearby Balboa Pavilion was rapidly becoming a popular vacation destination for bathers, boaters, and fishermen. The fire hall, a short block away, was comforting in its proximity to the businesses along Main Street. A large brass fire bell, visible in this photograph, was purchased for $44 and mounted on the roof. (Courtesy of NBFD archives.)

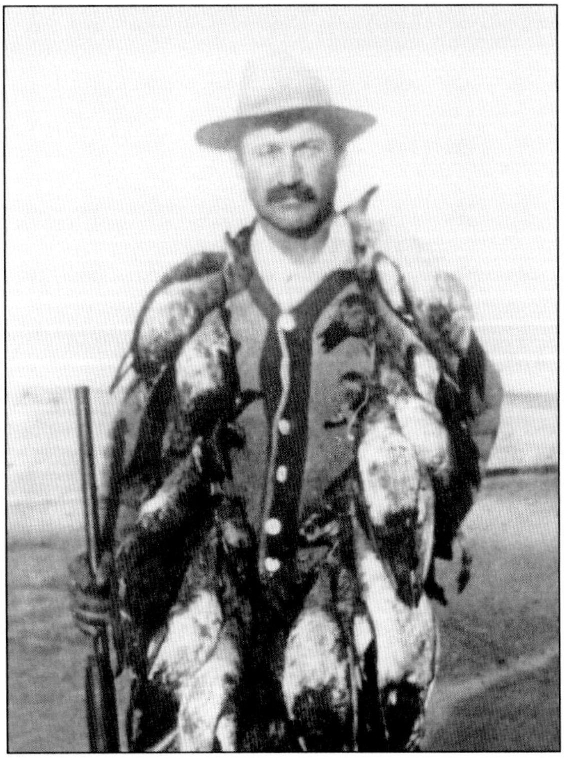

Rube Shafer was Newport's third volunteer fire chief, resigning in 1914. He was a well-known local fisherman. Shafer was best remembered for attempting to tow the schooner *Muriel* out to sea for use as a bait barge, only to see the ship run aground at the harbor entrance. The *Muriel* remained stranded there for five years. Between 1911 and 1927, seven men would serve as volunteer fire chief. The last volunteer chief, A. W. Jackson, served from 1921 to 1927. (Courtesy of NBFD archives.)

City of Newport Beach

INCORPORATED SEPTEMBER 1, 1906

NEWPORT BEACH, CALIFORNIA. Dec, 30th, 1915.

Fire Department Chief
Long Beach,
California.

Dear Sir:-

Will you please furnish me the following information, in connection with the proposed sale of your two small chemical trucks;

First; the cost of the trucks when new,

Second; their weight,

Third; the wheel base of each,

Trusting you will favor me with this information at your earliest convenience. I am.

Very truly yours,

H. C. Castle

H. C. C.-C

Fire Chief.

As 1915 drew to a close, fire chief Herbert "Bert" Castle wrote to the Long Beach Fire Department inquiring about two Rambler hose wagons available for sale. It had become apparent that Newport's hose reels were no longer an adequate defense against the threat of fire. Chief Castle astutely regarded these two motorized chemical wagons as a way to provide faster and more effective fire protection to both Balboa and Newport. They were large enough to carry several men, could carry up to 800 feet of fire hose and, most importantly, they could travel to both ends of the growing community faster than any other fire apparatus available at the time. (Courtesy of fire chief Steve Lewis.)

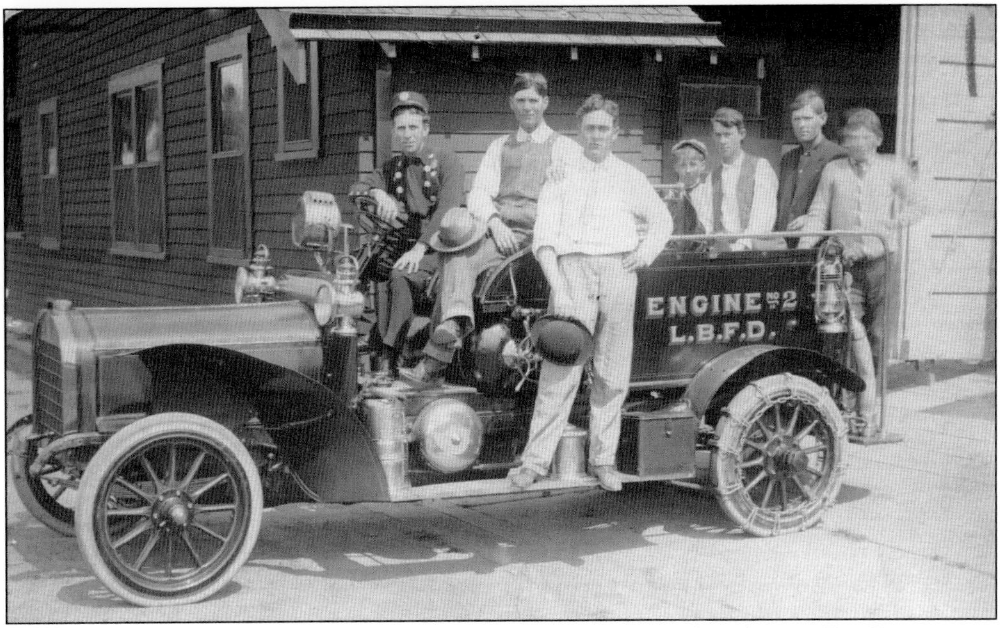

These are the Rambler chemical wagons sold to the Newport Beach Volunteer Fire Department in 1916. The Long Beach Fire Department purchased these two vehicles in 1907. They were the first motorized fire apparatus to reach the West Coast, and their delivery had made front-page news in Los Angeles. (Courtesy of Long Beach Fire Department Museum.)

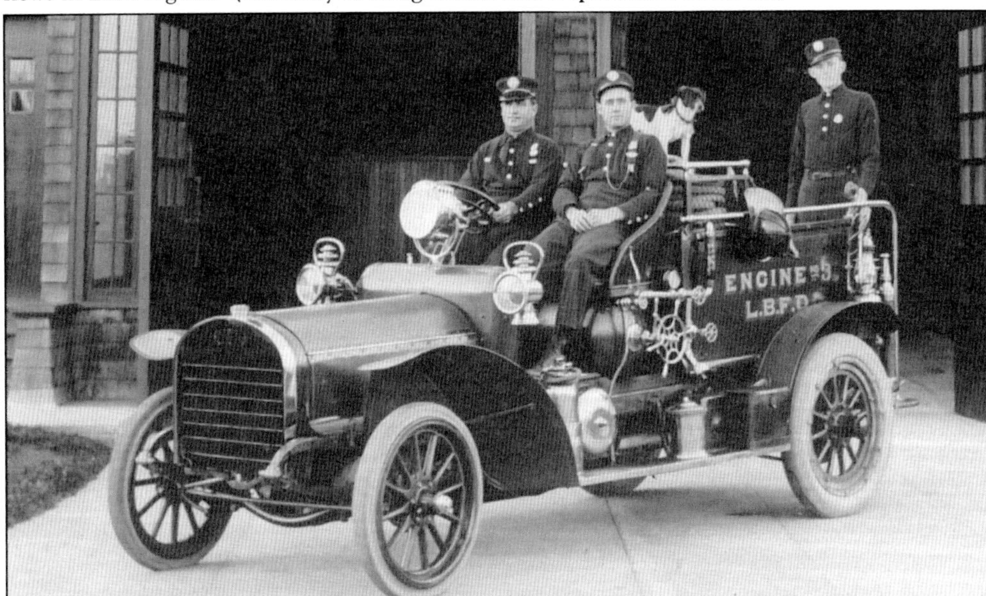

These Ramblers were unique in that they were designed to be parked on an incline ramp inside the fire station. At the sound of the alarm, the driver released a cable restraining the wagon. As the wagon rolled down the ramp, the driver could pop the vehicle into gear, jump-starting the engine. No hand cranking was necessary. Newport Village received their Rambler, designated No. 1, in March 1916. Balboa's Rambler No. 2 was delivered in June. (Courtesy of Long Beach Fire Department Museum.)

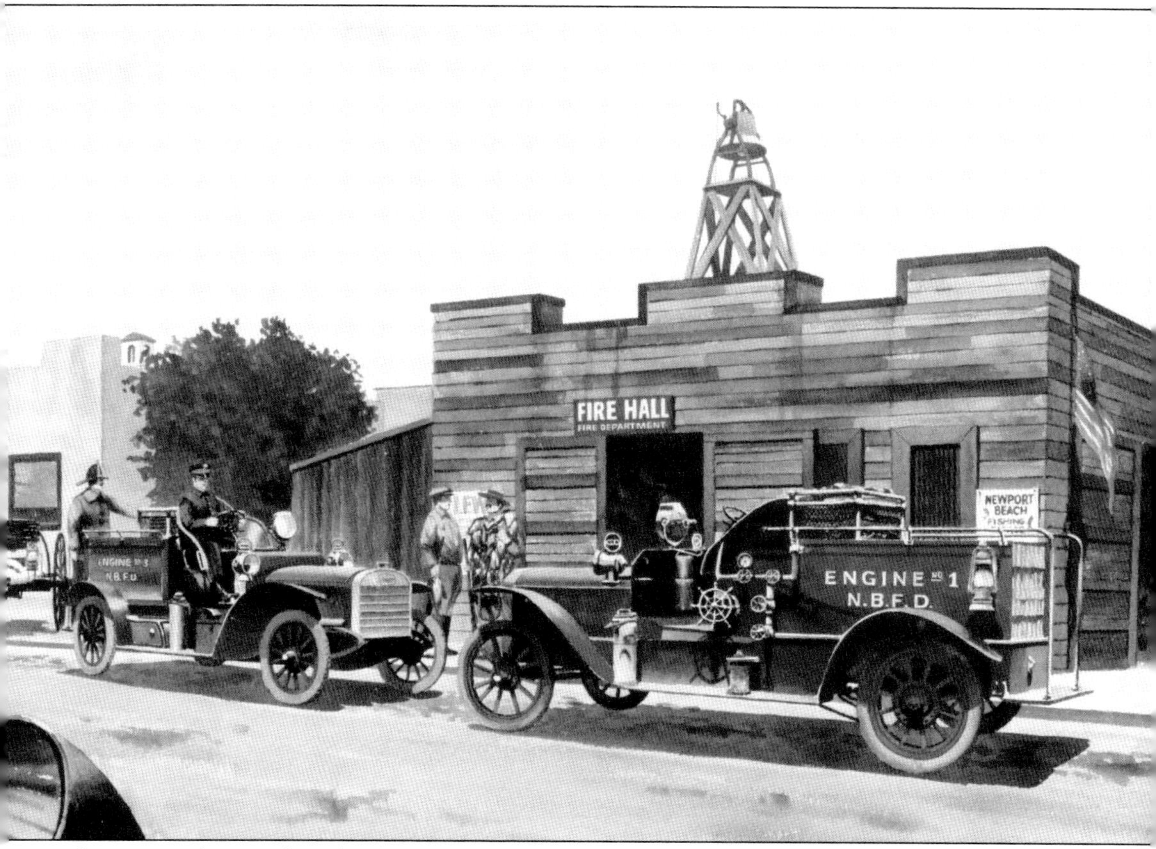

This watercolor was done for the Newport Beach Firefighters Association by local artist Ernest Toller. It is an accurate depiction of the fire hall on East Bay Street during the volunteer days. This hall was the center of the volunteer firefighter's universe. Meetings were held here. Mock trials provided weekly entertainment. Volunteers might be found "guilty or not guilty" for such transgressions as missing a meeting or a fire. Lack of an excuse often resulted in a guilty verdict, along with a small monetary fine. Department bylaws were passed and enacted. In 1918, $1 was collected from each member to put towards the purchase of an ambulance in France in support of the war effort. Bake sales were often held to raise money to pay for much needed firefighting equipment. After the meetings, volunteers moved down the street to the Balboa Hotel, or traveled to Sharps' Hotel in Newport Village, to dance or put on a "feed." These dinners usually ended with a sing-a-long fueled by good company and plentiful spirits. (Courtesy of NBFD archives.)

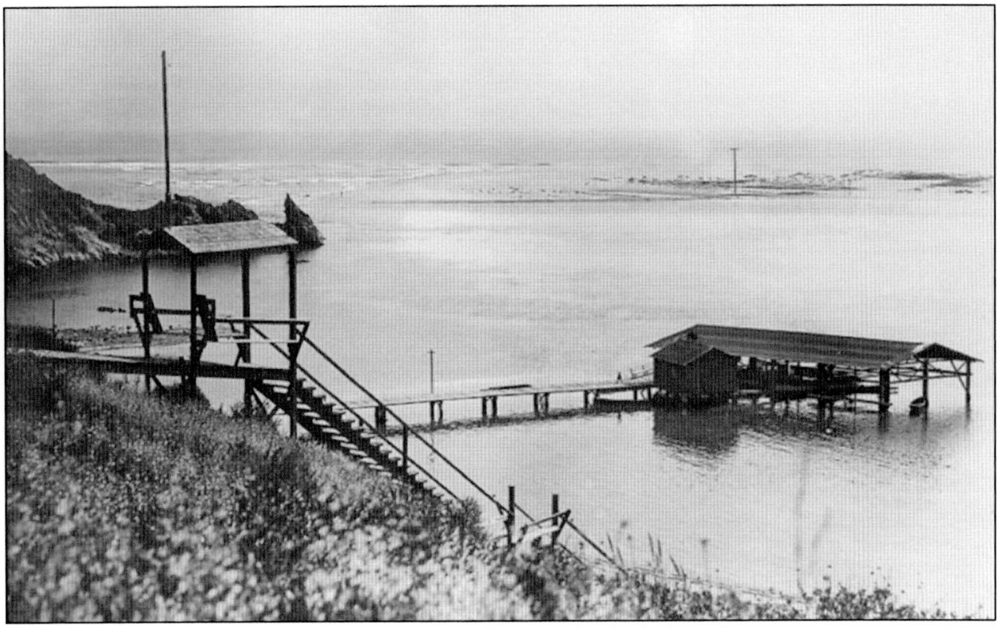

On a pleasant day in October 1913, Dr. Lowell Frost climbed aboard a primitive ferry and left the Hotel Del Mar wharf in Corona del Mar to buy groceries in Balboa Village, a short distance across the bay. He left his son and wife behind on the dock. While Dr. Frost was away, Lowell Jr. fell into the bay. Margaret Frost dove in to save her son. The strong ebb tide current that day swept them both into the breaking surf, where they drowned. (Author's collection.)

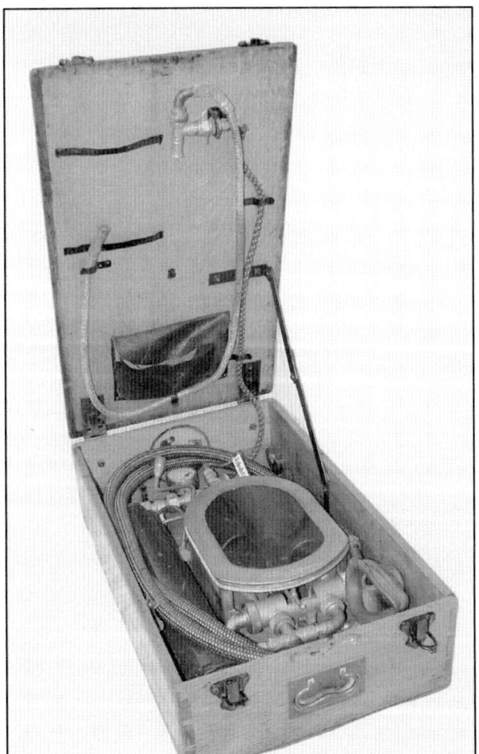

Dr. Frost, painfully aware of the lack of lifesaving equipment available to his family's would-be rescuers, presented the fire department with a Draeger Pulmotor in memory of his wife and son. The pulmotor was the latest in mechanical resuscitation equipment. It was placed on Rambler No. 2 and the Frost Life Saving Corps was organized to utilize it. Newport's first emergency medical care service was inaugurated. (Courtesy of NBFD archives.)

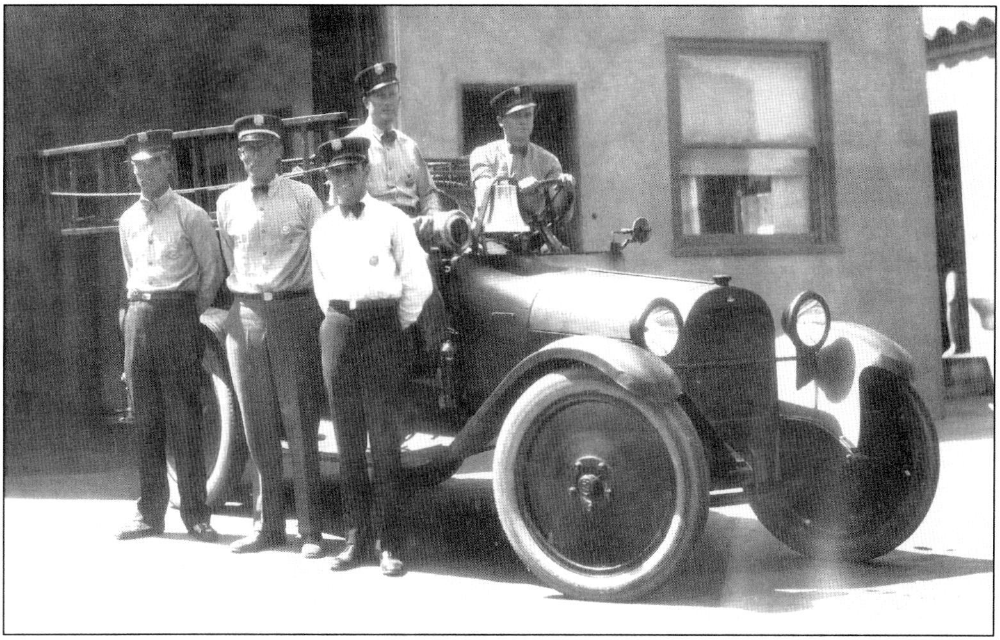

By 1920, Rambler No. 2 was on its last legs. A Dodge truck chassis was purchased and the Rambler's hose compartment, chemical tank, and equipment were transferred onto this new vehicle. Officially it was now called Engine No. 2, but it was commonly referred to as the "Squad." (Courtesy of NBFD archives.)

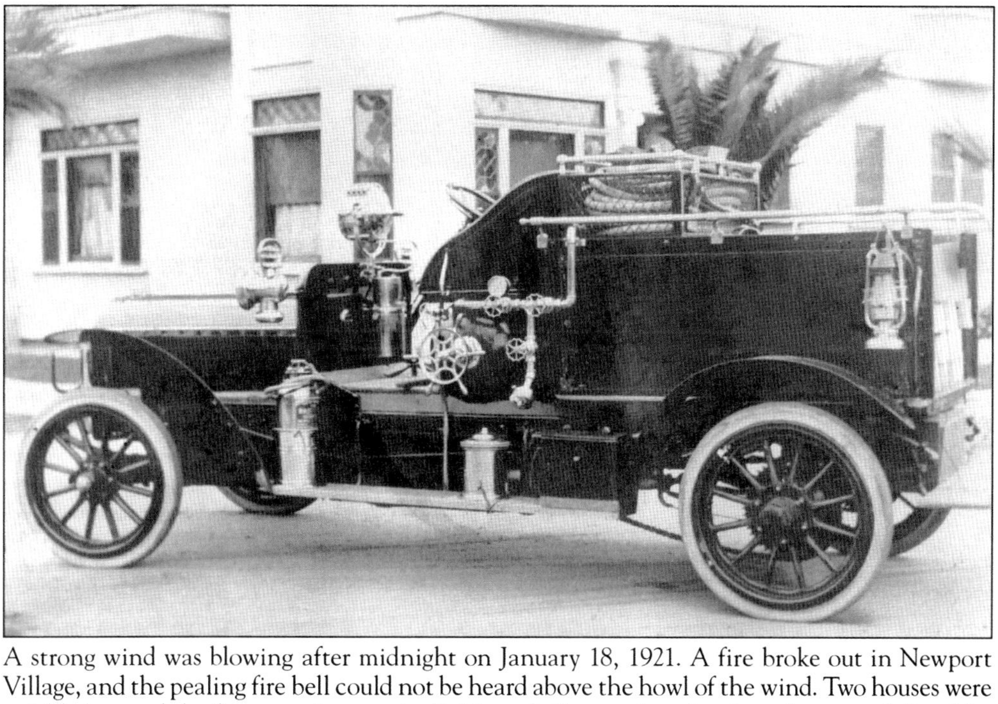

A strong wind was blowing after midnight on January 18, 1921. A fire broke out in Newport Village, and the pealing fire bell could not be heard above the howl of the wind. Two houses were quickly lost and the fire spread to nearby Robinson's Garage, burning it to the ground. Rambler No. 1, stationed inside the garage since 1916, was destroyed in the blaze, along with the steam whistle used to alert volunteers in Newport Village. (Courtesy of NBFD archives.)

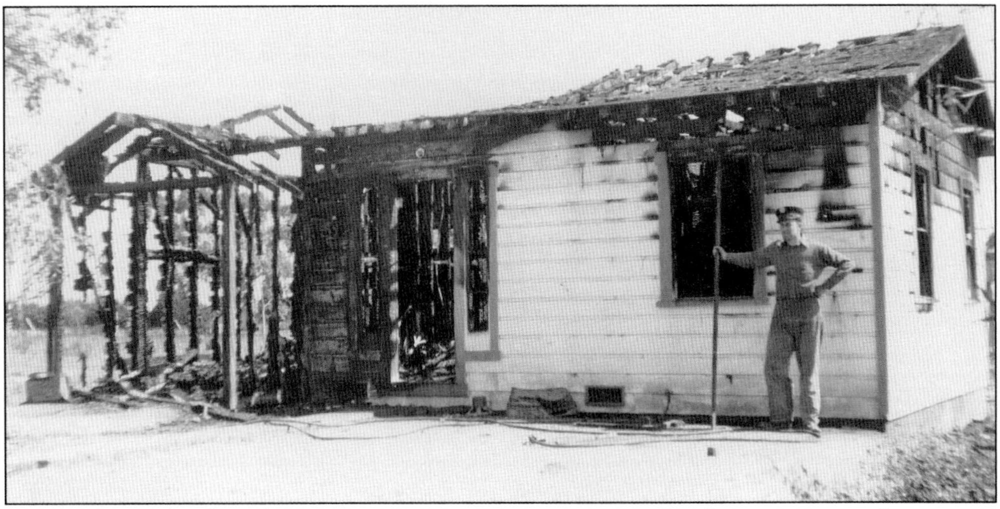

The Newport Beach Volunteer Fire Department was now reduced to one piece of motorized fire apparatus. Throughout the early 1920s, Chief Jackson worked hard to convince the city trustees to improve fire protection in the city, all to no avail. Bond elections to build a new station continually failed. There was little money available for new hose and equipment. Despite this lack of support, volunteer firefighters continued to respond to all emergencies with spirited enthusiasm. (Courtesy of NBFD archives.)

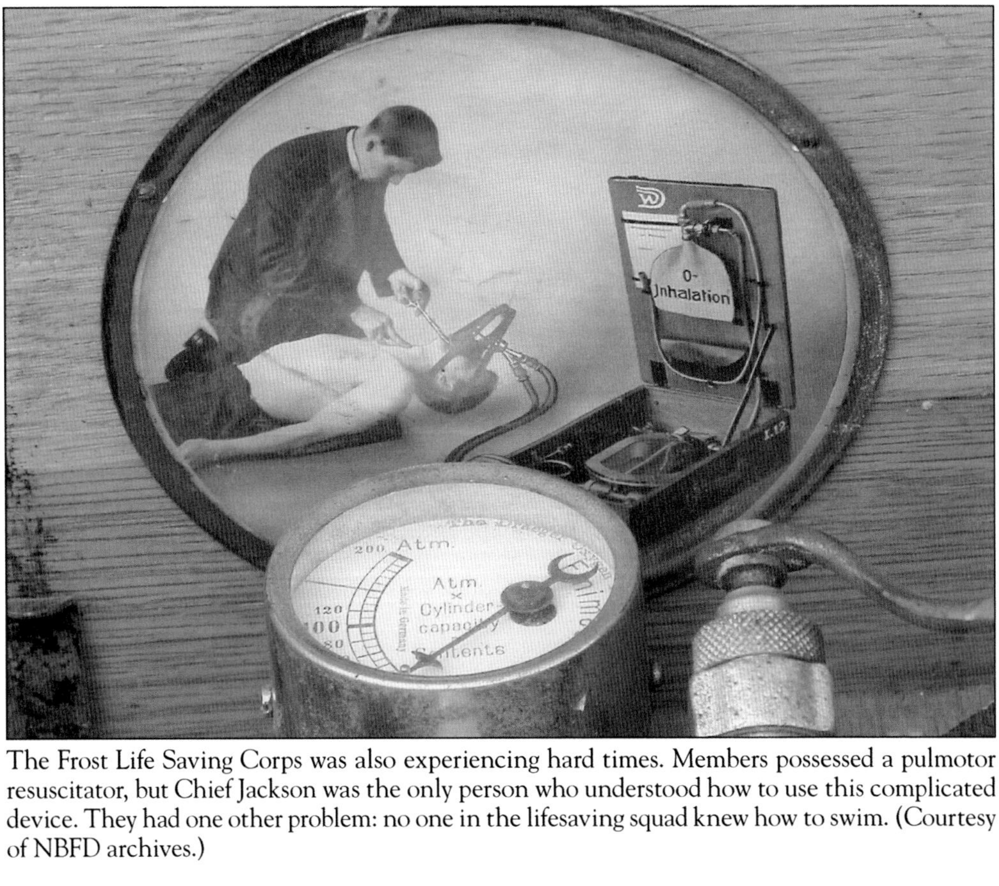

The Frost Life Saving Corps was also experiencing hard times. Members possessed a pulmotor resuscitator, but Chief Jackson was the only person who understood how to use this complicated device. They had one other problem: no one in the lifesaving squad knew how to swim. (Courtesy of NBFD archives.)

Two

A Professional Department Is Born

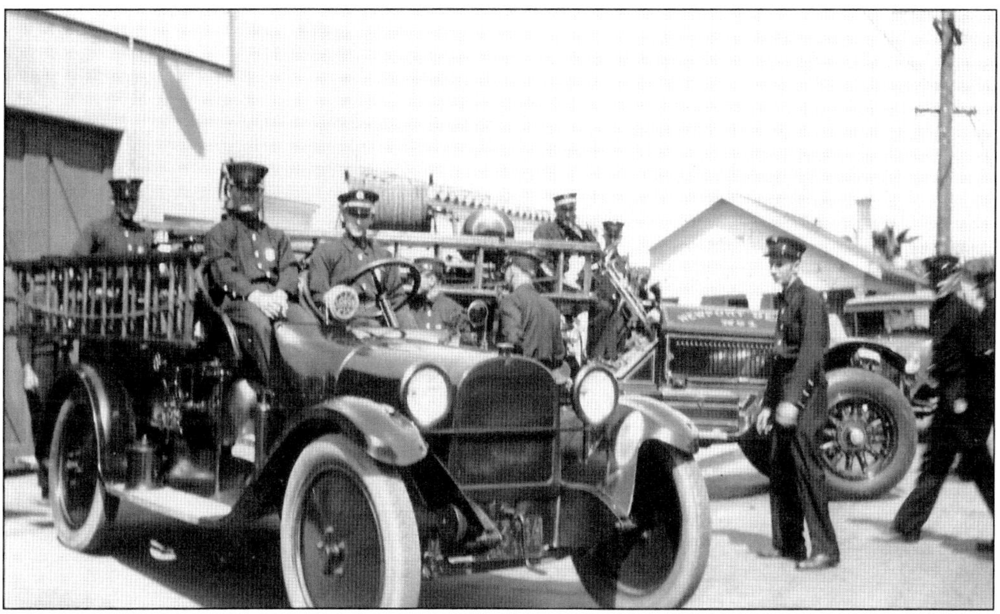

By 1927, it was becoming increasingly apparent that a major improvement in fire protection was necessary. The city was growing rapidly. Balboa Island and Corona del Mar were clamoring for better fire protection. Chief Jackson's persistent attempts to fund a new fire station and fire engine through bond elections had repeatedly failed. Frustrated firemen were doing the best they could with a bare minimum of equipment and support. In December 1926, the board of underwriters surveyed fire protection in Newport Beach for insurance purposes. Their final report was extremely unfavorable. Goaded into action, the City of Newport Beach ordered an American LaFrance fire pumper. Ordinance No. 315 was drafted, mandating a complete reorganization of the fire department. The volunteer era was over. (Courtesy of NBFD archives.)

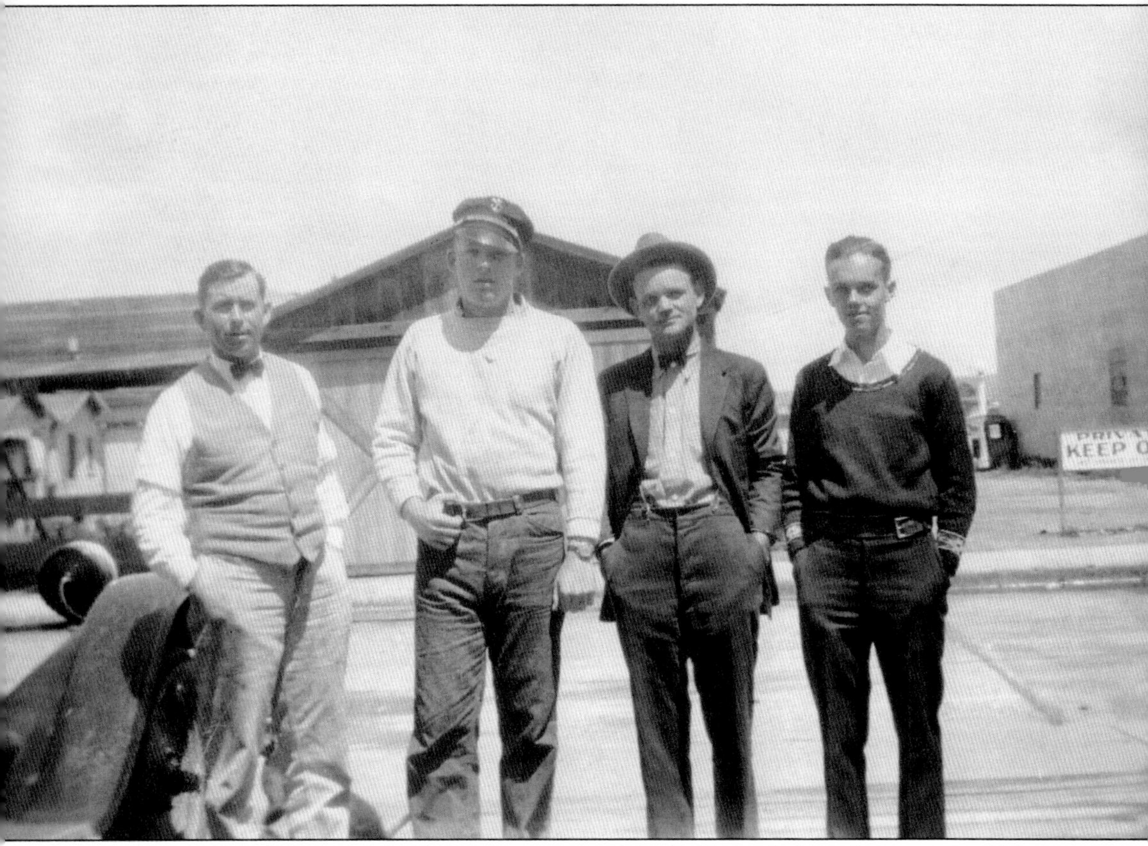

Ordinance No. 315 was ratified by city trustees on March 28, 1927. A professional, paid Newport Beach Fire Department was now a reality. The first full-time fire personnel were hired. Frank Crocker was appointed fire chief at a salary of $200 per month, and Clinton Dickey was hired as driver for $150 per month. In this photograph, newly appointed fire chief Frank Crocker stands on the far left. Crocker's first act was to undertake a complete reorganization of the fire department and a reevaluation of its efficiency. A youthful Jan Briscoe, destined to become Newport's second fire chief, stands on the right. (Courtesy of NBFD archives.)

In 1927, the air whistle that sounded the fire alarm required the driver to activate it by hand before leaving the station. It could take up to three minutes to perform this task, depending upon the number of alarm blasts required. This Gamewell Alarm Transmitter was purchased secondhand from the Portland, Oregon, fire department to automate this procedure. Response time out of the station was reduced to 15 seconds. (Author's collection.)

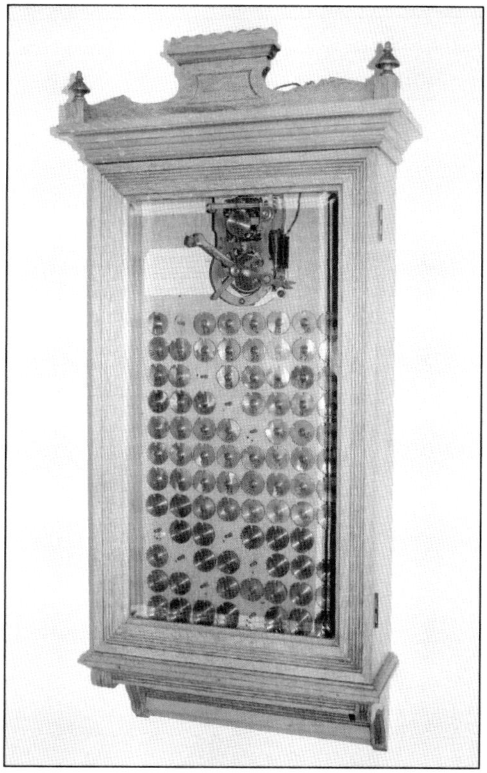

In March 1927, Newport Beach took delivery of an American LaFrance Type 75 fire engine made in Elmira, New York. It cost $14,000. It was Newport's first piece of fire apparatus to incorporate a fire pump. Rated at 750 gallons per minute (GPM), this engine could supply water through multiple hose lines at high pressure. (Courtesy of NBFD archives.)

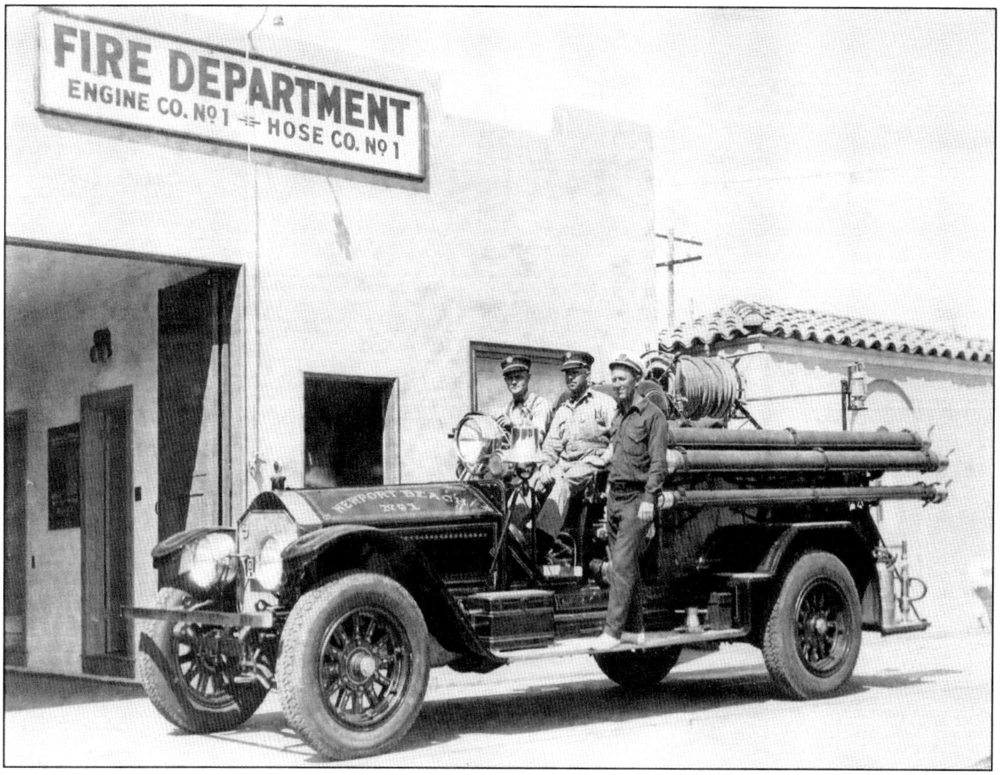

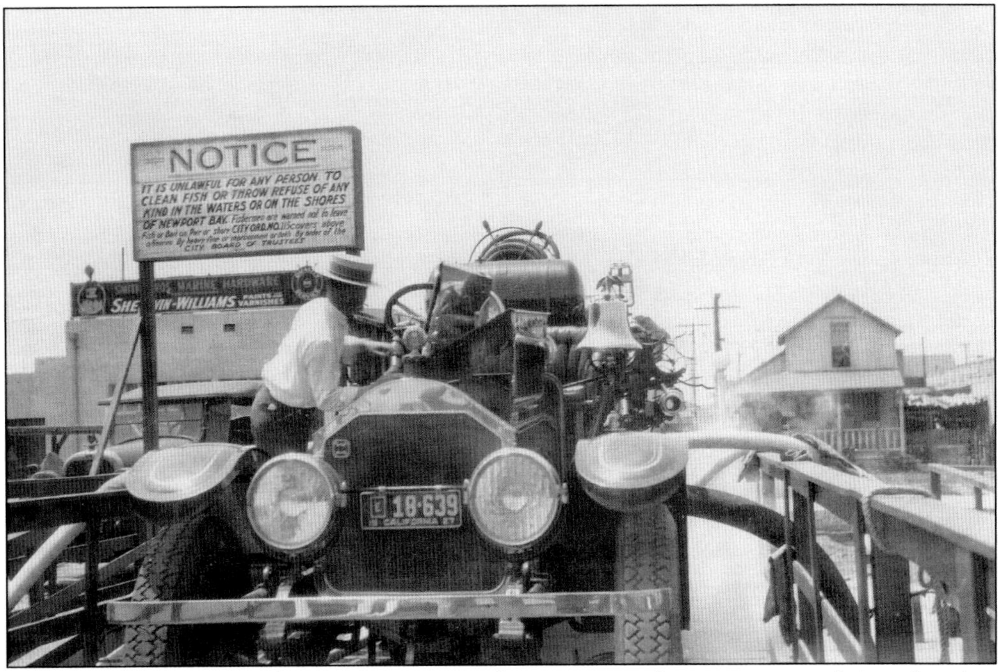

Fire engines were required to meet certain performance specifications prior to being placed in service. This photograph shows Newport Engine No. 1 undergoing its Board of Underwriter's acceptance test. These tests were conducted to insure that the engine's pump could perform as specified by the manufacturer. (Courtesy of NBFD archives.)

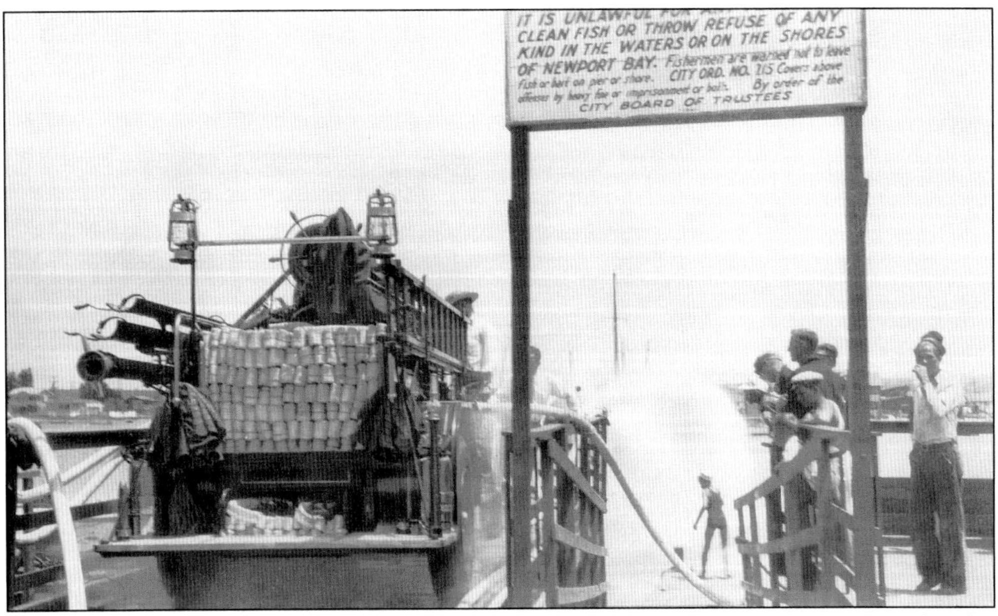

Several children managed to find a way to turn Engine No. 1's acceptance test into a fun day at the beach. Note the new hose neatly stacked in the hose bed. Engine No. 1's fire hose would never look this clean again. (Courtesy of NBFD archives.)

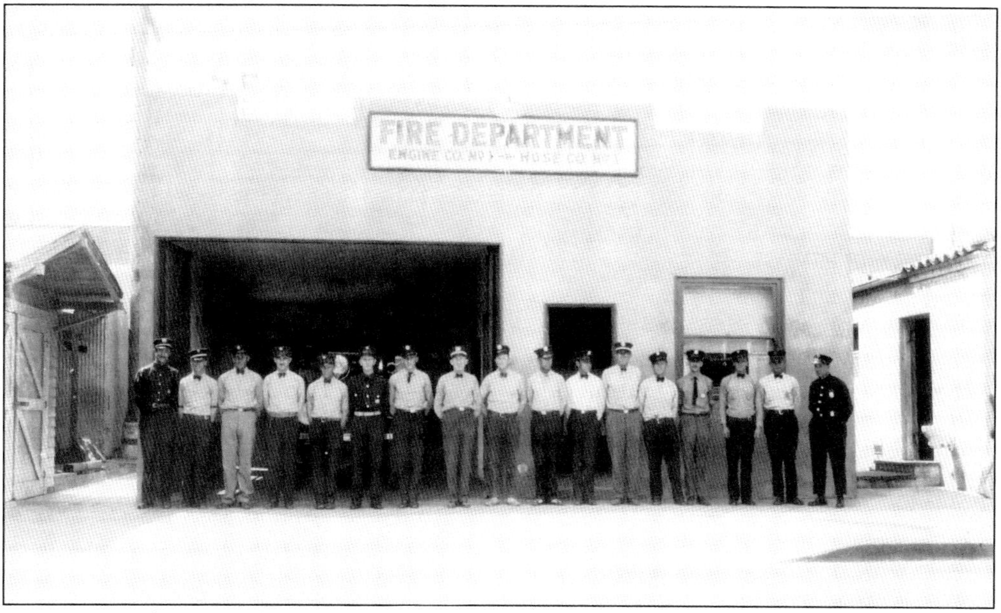

Newport Beach's original fire hall was abandoned and moved to another location early in 1927. A new fire station and public restroom were erected on the site. Chief Frank Crocker stands on the far right, and driver Clinton Dickey stands second from the right. The rest of the men in this photograph are paid-call volunteers. Several would eventually move into salaried positions. (Courtesy of NBFD archives.)

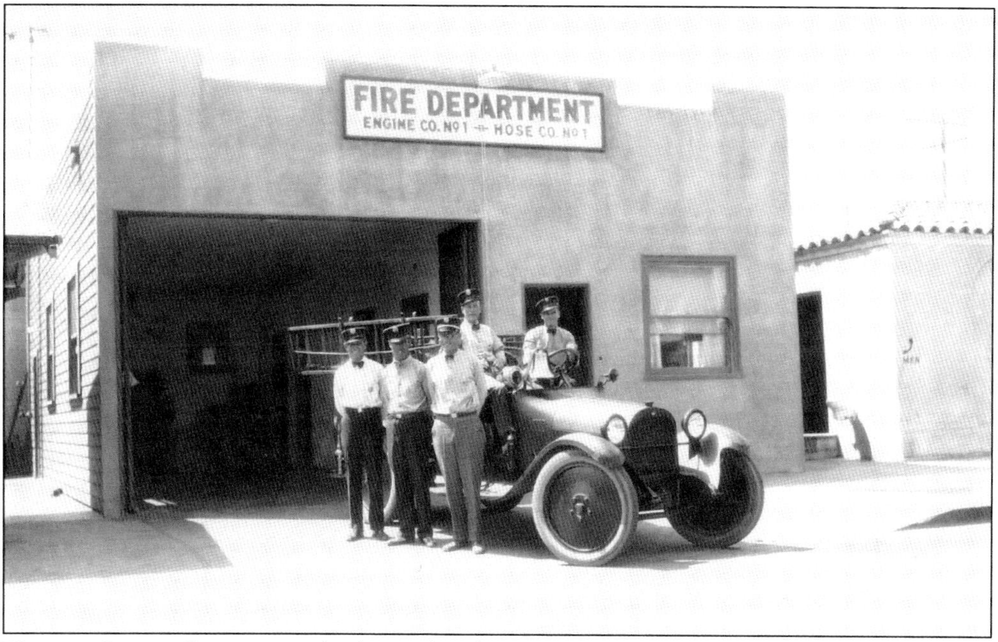

The new Station One was nearly identical in design to the old wood fire hall formerly occupying the site. Politics being what they are, the city trustees added a public restroom to the appropriation request to justify the funding of the fire station. Rather than ask city residents to choose between two necessities, packaging fire protection and the personal comfort of visitors together made voting "yes" for the expenditure easier to accept. (Courtesy of NBFD archives.)

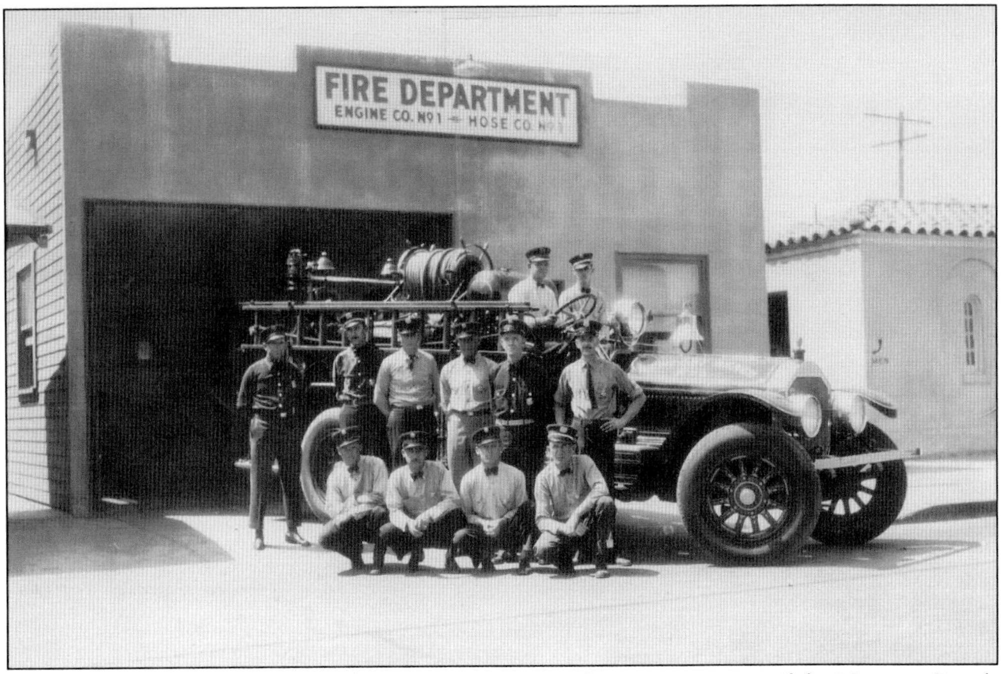

This series of photographs were taken to commemorate the reorganization of the Newport Beach Fire Department and the opening of Station One in July 1927. Copies were made available to the firefighters as souvenirs of the occasion. (Courtesy of NBFD archives.)

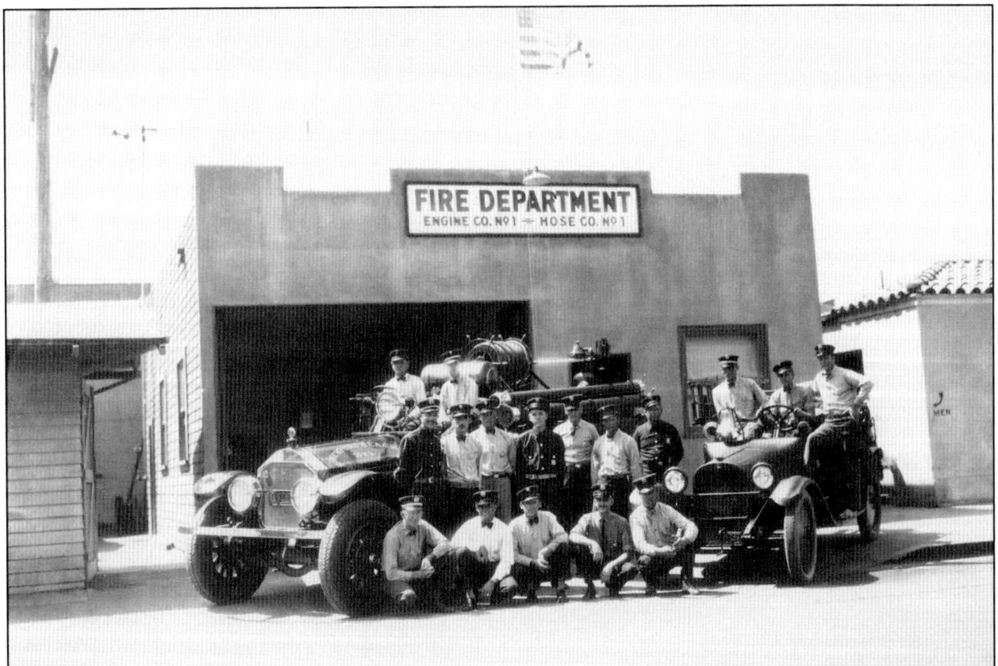

The entire Newport Beach Fire Department is pictured here. The Dodge squad on the right would soon be temporarily relocated to Balboa Island, where it would remain until 1931. (Courtesy of NBFD archives.)

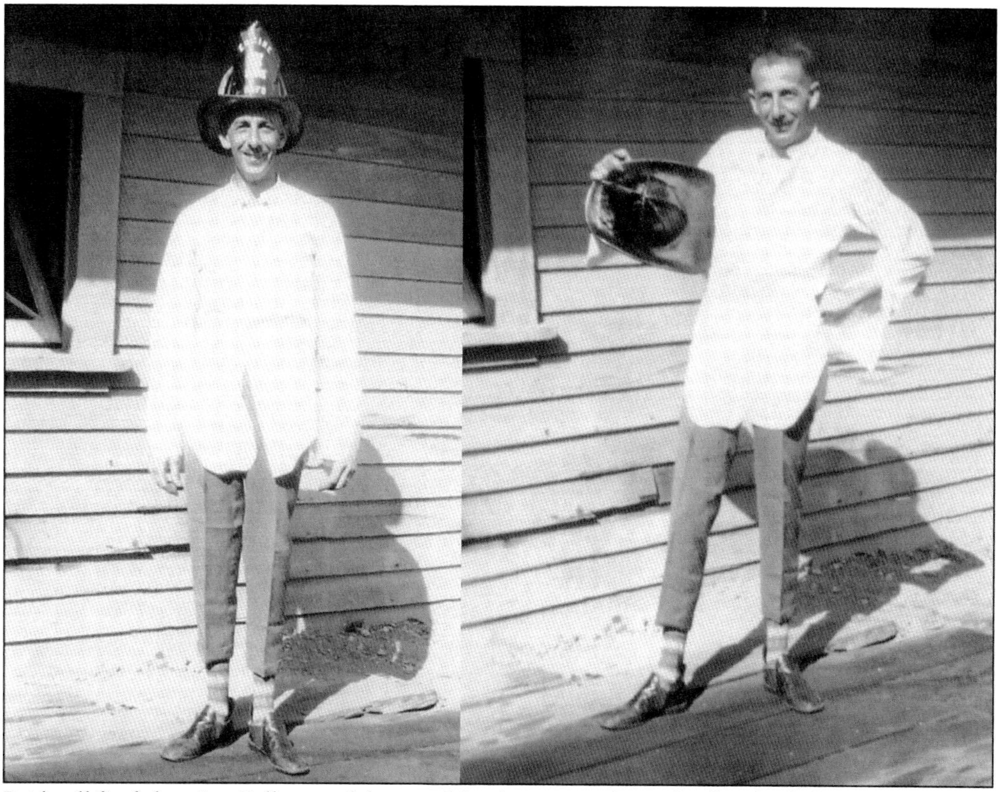

Paid-call firefighter Lee Pellett models one of the department's new leather fire helmets. Six helmets were purchased and suspended from the grab bar at the rear of Engine No. 1. Firefighters quick enough to jump on the tailboard of the engine had a protective helmet to wear upon arrival at a fire. Pellett's socks and pants were not department issue. (Courtesy of Don Crocker.)

This is the only known surviving Newport Beach fire helmet from 1927. It was made by Cairns and Brothers of New York City. These helmets were essential when fighting fire. They protected a firefighter's head from injury and deflected intense heat. (Author's collection; helmet in the Sherman Library collection.)

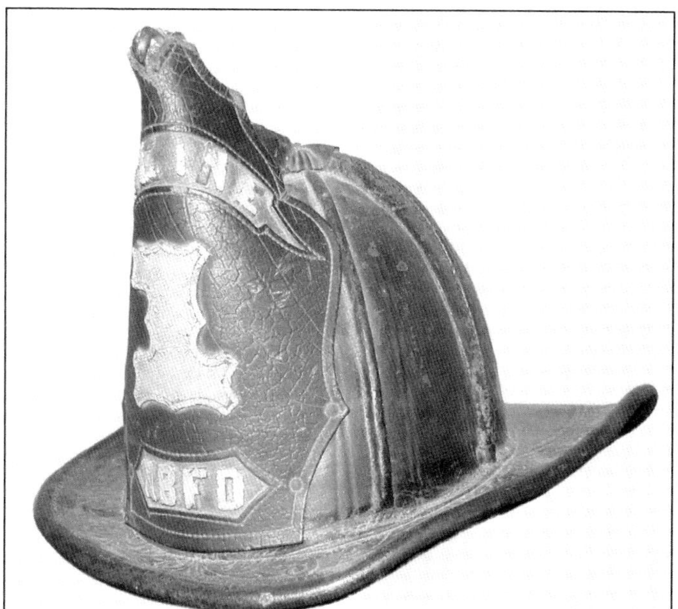

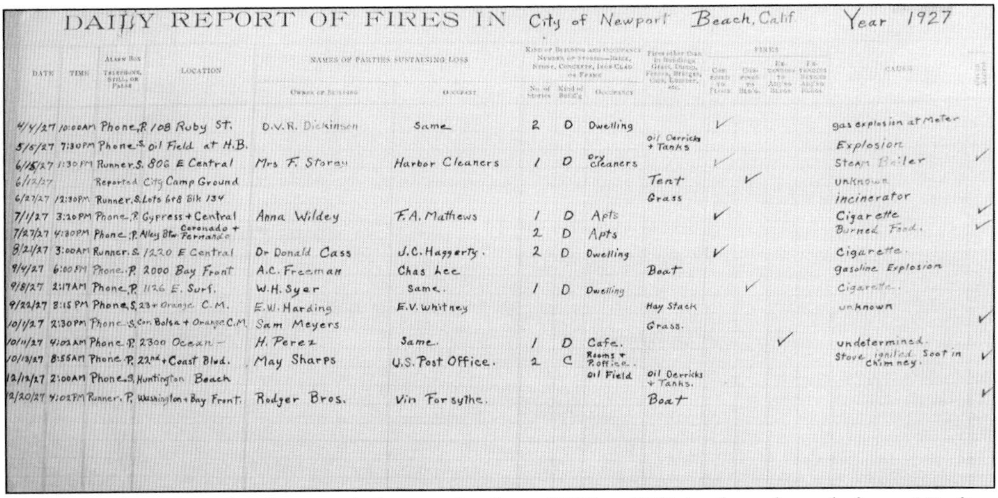

DAILY REPORT OF FIRES IN City of Newport Beach, Calif. Year 1927

Date	Time	Alarm Box: Telephone, Still, or Pass	Location	Owner of Building	Occupant	No. of Stories	Kind of Bldg	Occupancy	Fires other than in Buildings	Confined to Floor	Confined to Bldg	Extending to Adj'g Bldgs	Extending beyond Adj'g Bldgs	Cause	
4/4/27	10:00AM	Phone-P.	108 Ruby St.	D.V.R. Dickinson	Same	2	D	Dwelling		✓				gas explosion at Meter	
5/5/27	7:30PM	Phone-S.	oil Field at H.B.						oil Derricks + Tanks					Explosion	
6/15/27	1:30PM	Runner.S.	806 E Central	Mrs F. Storay	Harbor Cleaners	1	D	Dry Cleaners		✓				Steam Boiler	✓
6/12/27		Reported	City Camp Ground						Tent		✓			unknown	
6/27/27	12:30PM	Runner.S.	Lots 6+8 Blk 134						Grass					incinerator	
7/1/27	3:20PM	Phone-P.	Cypress + Central	Anna Wildey	F.A. Mathews	1	D	Apts		✓				Cigarette	✓
7/27/27	4:30PM	Phone-P.	Alley Btw. Coronado + Fernando			2	D	Apts						Burned Food.	✓
8/21/27	3:00AM	Runner.S.	1220 E Central	Dr Donald Cass	J.C. Haggerty.	2	D	Dwelling		✓				Cigarette.	
9/4/27	6:00PM	Phone-P.	2000 Bay Front	A.C. Freeman	Chas Lee				Boat					gasoline Explosion	
9/8/27	2:17AM	Phone-P.	1126 E. Surf.	W.H. Syer	Same.	1	D	Dwelling			✓			Cigarette.	
9/22/27	8:15 PM	Phone-S.	23 + Orange C.M.	E.W. Harding	E.V. Whitney				Hay Stack					unknown	
10/1/27	2:30PM	Phone-S.	Cor. Bolsa + Orange C.M.	Sam Meyers					Grass.						✓
10/11/27	4:02AM	Phone-P.	2300 Ocean –	H. Perez	Same.	1	D	Cafe.				✓		undetermined.	
10/13/27	8:55AM	Phone-P.	22nd + Coast Blvd.	May Sharps	U.S. Post Office.	2	C	Rooms + P. office.						Stove ignited Soot in Chimney.	✓
12/12/27	2:00AM	Phone-S.	Huntington Beach					oil Field	oil Derricks + Tanks.						
12/20/27	4:02PM	Runner.P.	Washington + Bay Front.	Rodger Bros.	Vin Forsythe.				Boat						✓

Chief Crocker meticulously recorded every fire from 1927 to 1952 in three large ledgers. His first entry was dated April 4, 1927. The ledgers listed the dates, times, and locations of the fires, the actions taken, and a host of additional details. All entries were handwritten until the late 1940s, when he switched to a typewriter. (Courtesy of NBFD archives.)

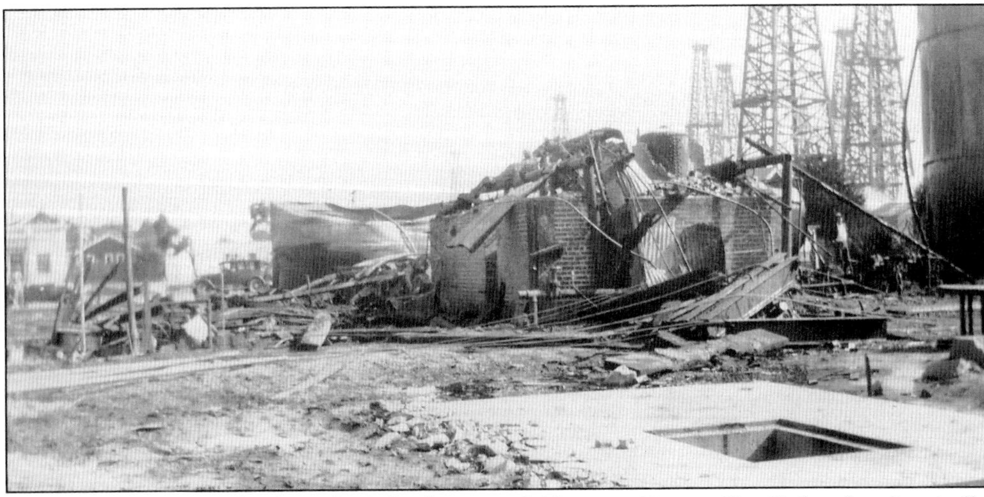

The second entry in Chief Crocker's fire log records Newport Engine No. 1's first fire. Ironically, this fire was in nearby Huntington Beach, where an explosion rocked an oil field, killing one person. Engine No. 1 battled the fire for three hours and was instrumental in extinguishing the blaze. (Courtesy of NBFD archives.)

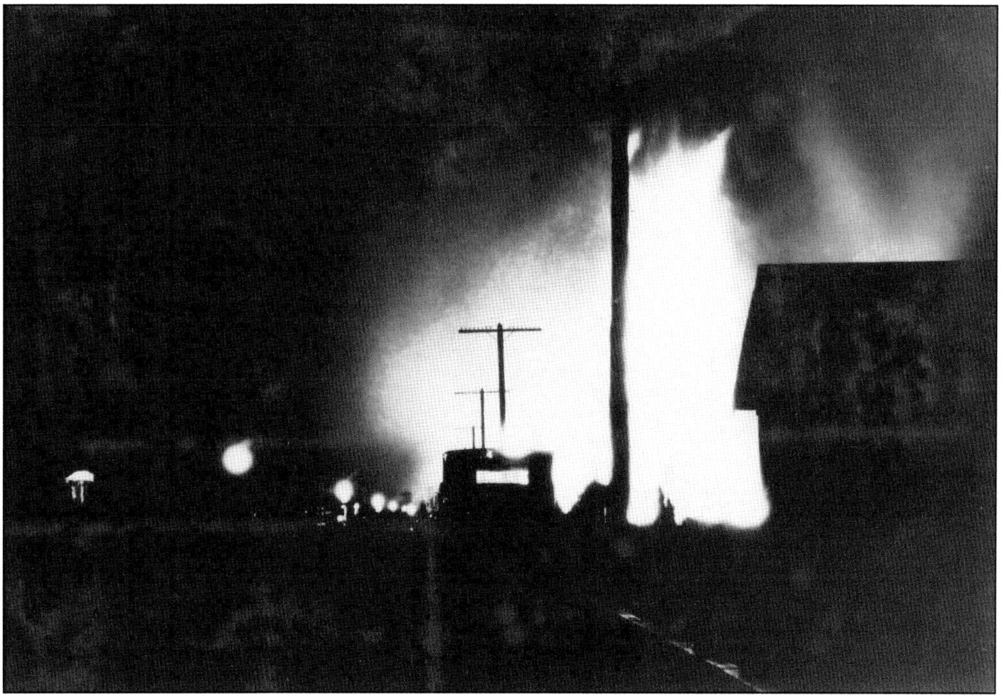

Engine No. 1's first big trial by fire in Newport Beach was another oil field fire. The Orange County Refining Company operated in West Newport at Forty-third Street and Balboa Avenue. Fire broke out on a January night in 1928. Two oil rigs, a warehouse, and 1,400 barrels of asphalt were destroyed. (Courtesy of NBFD archives.)

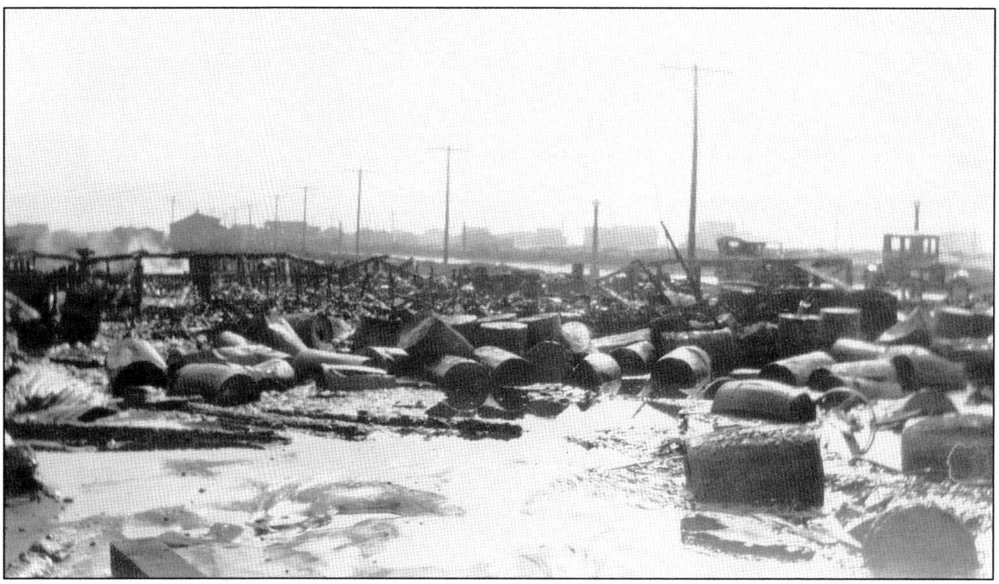

Upon his retirement, Chief Briscoe reminisced that this refinery site was a constant headache early in his career. Many fire alarms originated there. In 1929, the wells were capped and a community of beach houses eventually removed all trace of the oil field. Fifty-six years later, in 1985, one of those caps failed. The resulting fire and oil spill took firefighters hours to control. (Courtesy of NBFD archives.)

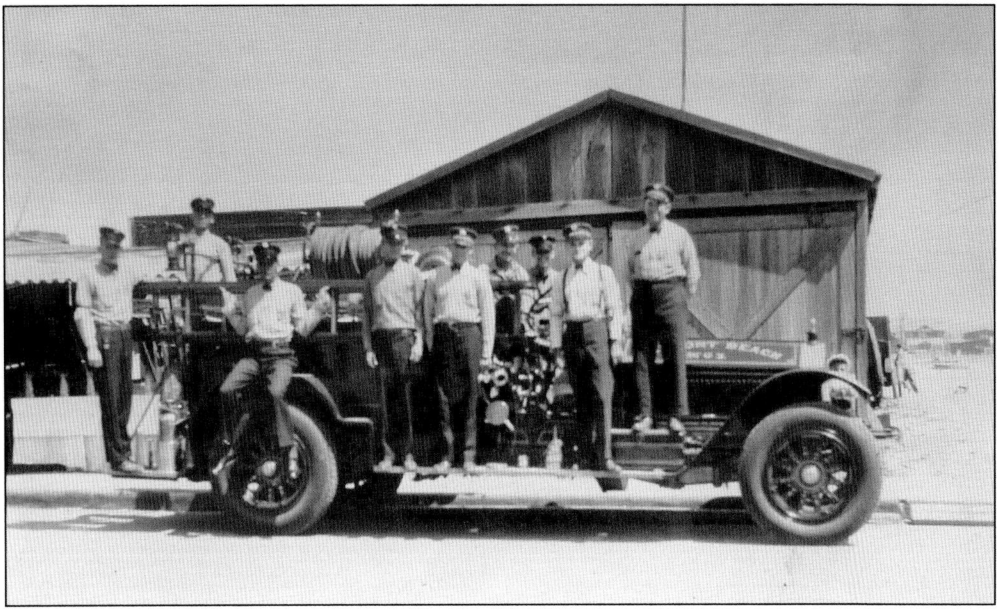

The Boys of 1927 demonstrate how many firefighters can fit on an American LaFrance fire engine. In practice, the LaFrance rarely responded to an emergency call with more than three firefighters. (Courtesy of NBFD archives.)

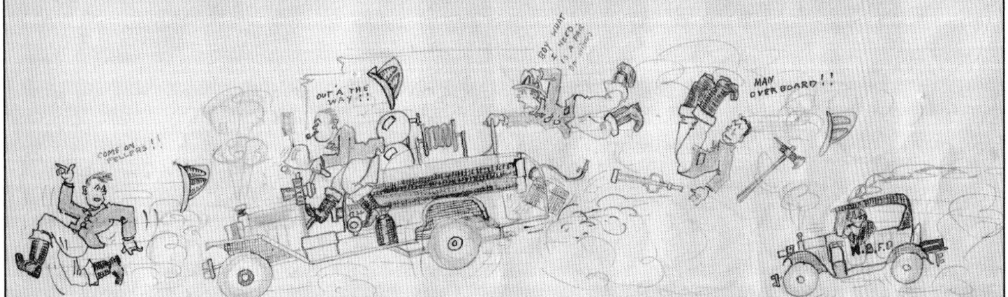

This light-hearted depiction of the Newport Beach Fire Department responding to a fire was drawn in 1929. Self-deprecating humor was, and still is, a staple of firehouse life. (Courtesy of NBFD archives.)

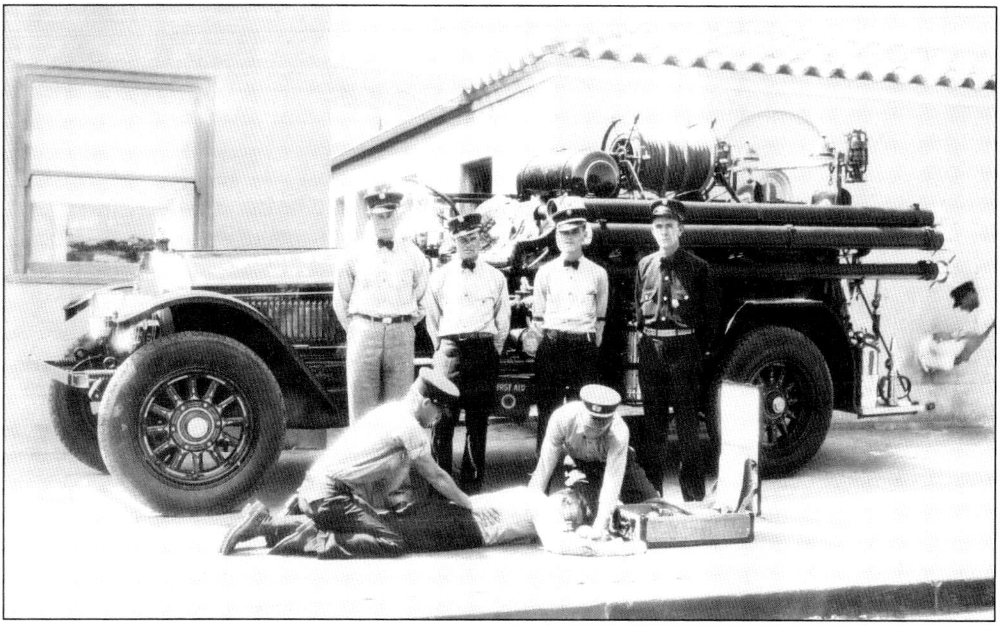

A rescue squad was created in September 1927 for the express purpose of providing resuscitation in the case of drowning, smoke inhalation, or medical emergency. Several members of the Newport Beach Fire Department were given specialized training by the Long Beach Fire Department. In this photograph, they demonstrate their newly learned resuscitation techniques. (Courtesy of NBFD archives.)

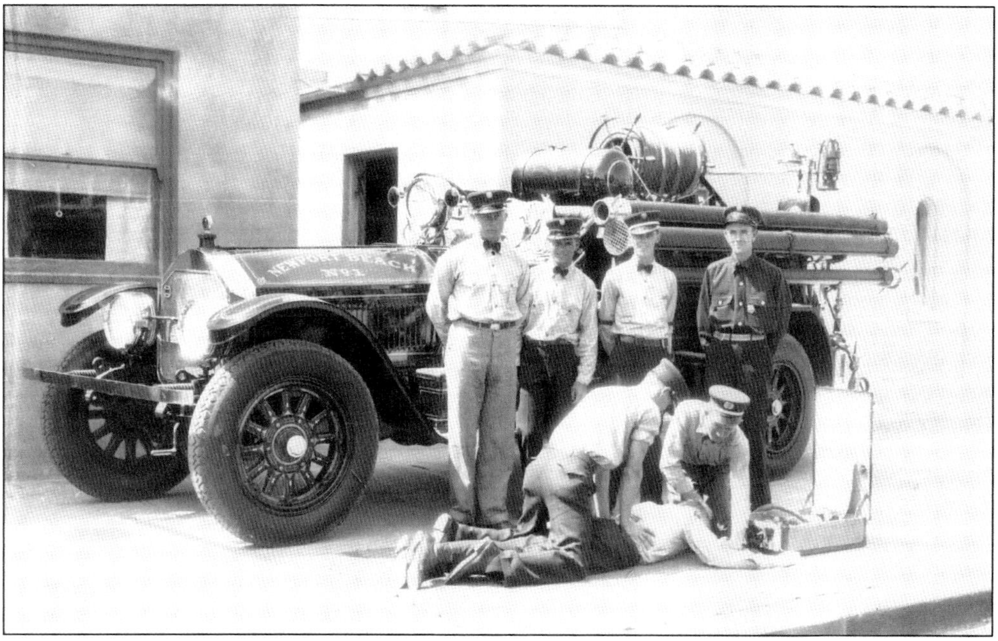

The lifesaving techniques being demonstrated by the rescue squad were first espoused in the Boy Scout Handbook of 1911. The Boy Scouts based their method on an early form of artificial respiration known as the Silvester Method. The rescue squad is shown applying their lifesaving skills in conjunction with a newly purchased E & J resuscitator. (Courtesy of Ray Shannon family.)

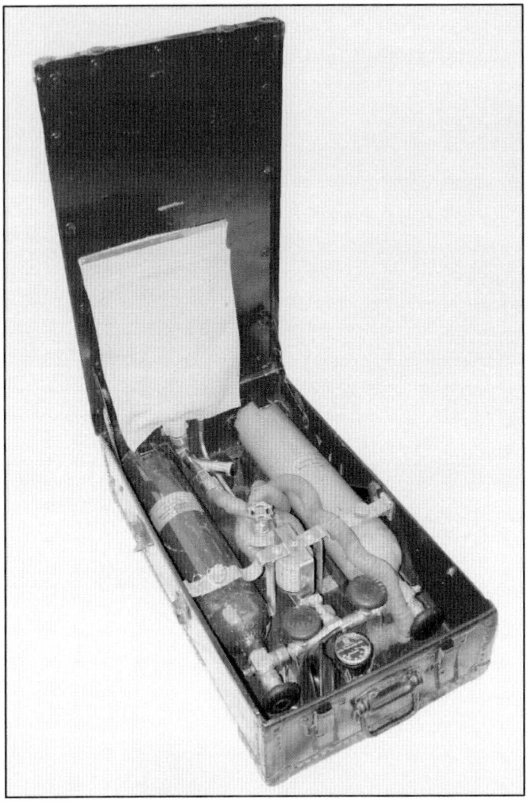

This is the E & J resuscitator visible in the photographs of the rescue squad. It was purchased for the princely sum of $350, a price tag nearly equal to the cost of a new Ford or Chevy automobile at the time. Logbook entries indicate that the E & J was often loaned to fire personnel when family members were hospitalized. (Courtesy of NBFD archives.)

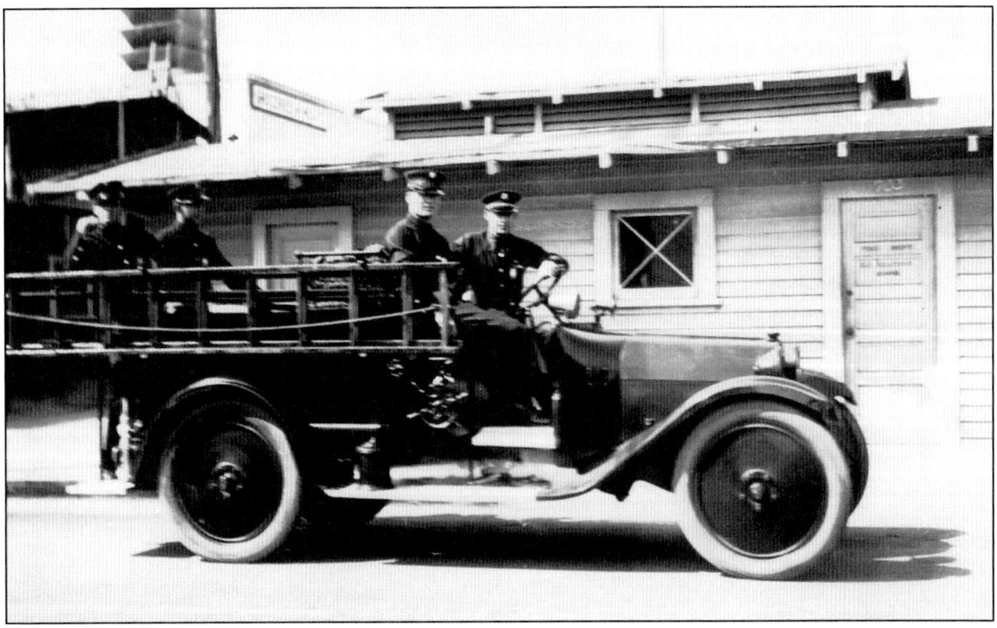

The E & J proved so effective that Chief Crocker purchased a high-pressure oxygen pump that would enable the department to refill its own oxygen bottles, saving both time and expense. The E & J was carried on the Dodge Squad, shown here parked in front of lifeguard headquarters. (Courtesy of NBFD archives.)

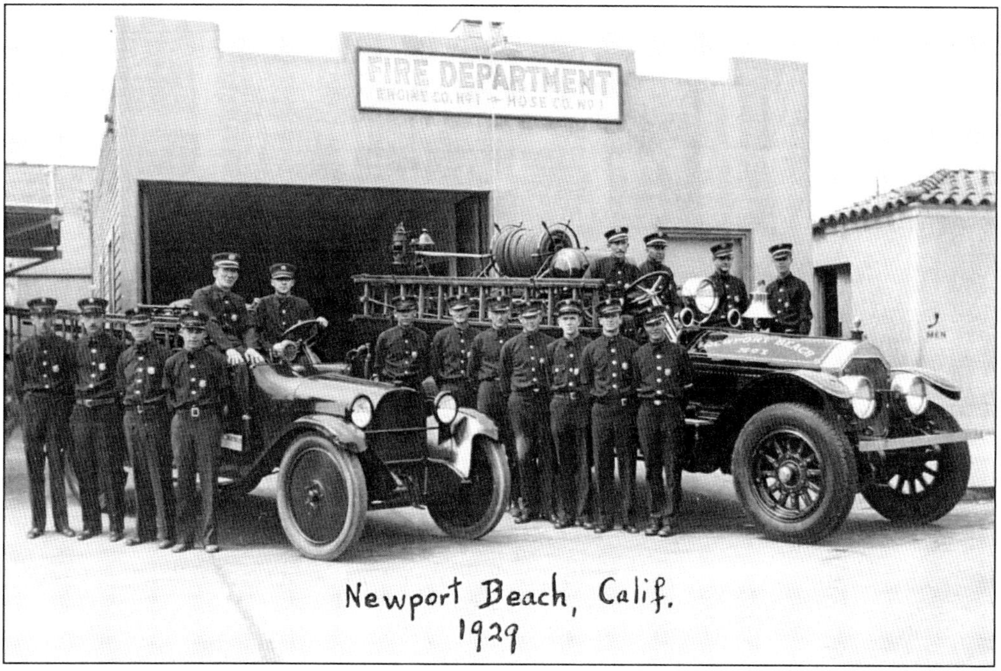

What a difference a year or two makes. New uniforms were issued in 1929. Each member also received an individually numbered fire badge. Chief Crocker was determined to make the Newport Beach Fire Department into one of Orange County's finest, and his efforts were beginning to pay off. (Courtesy of NBFD archives.)

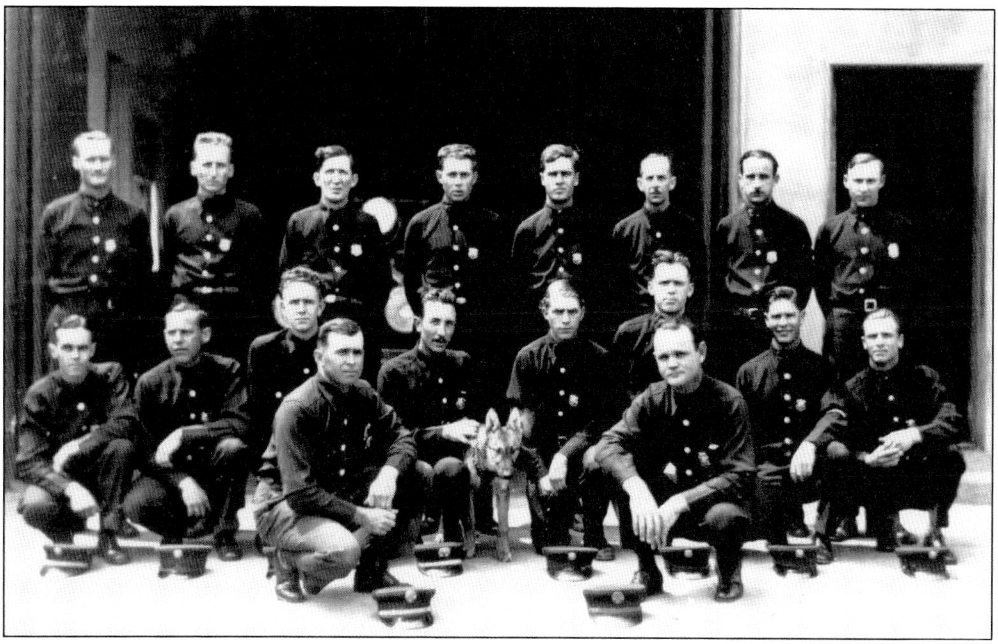

Frank Crocker and Clinton Dickey, kneeling in front, remained the only salaried firefighters in 1929. The paid-call firefighters formed the Newport Beach Fireman's Club in 1927. Fred Barnett, standing fourth from the left, was elected president. The Fireman's Club was a precursor to the present-day Newport Beach Firefighter's Association. (Courtesy of NBFD archives.)

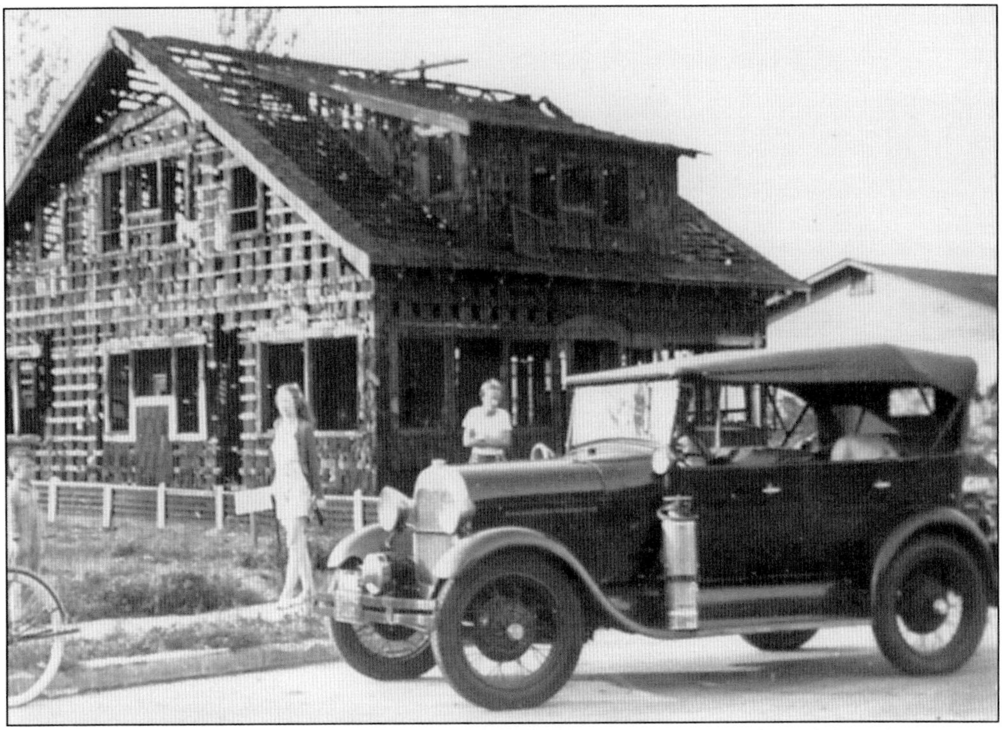

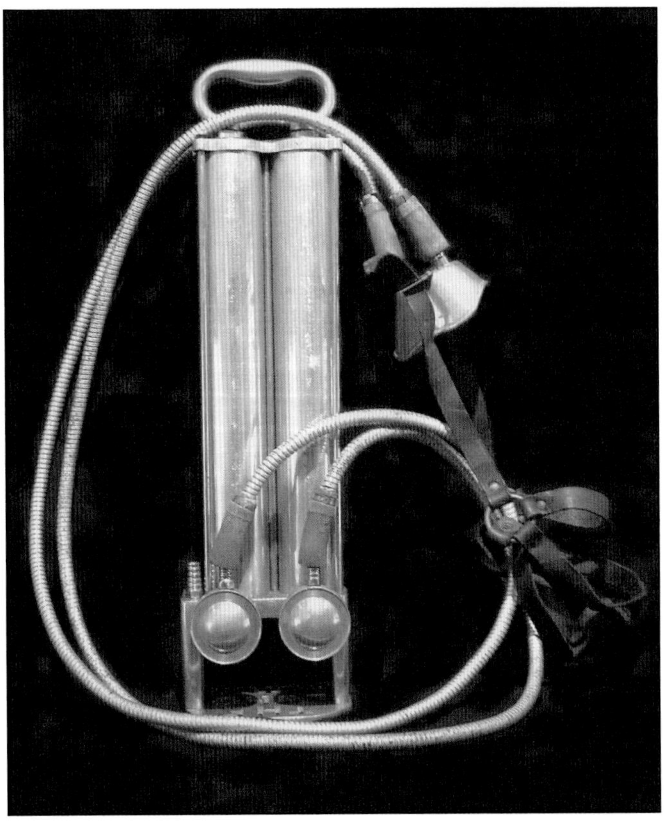

A Ford Model A Phaeton was purchased in May 1929. It was equipped with a fire extinguisher and a resuscitator. This vehicle gave the fire department more versatility when it came to handling lifesaving emergencies and small fires. (Courtesy of NBFD archives.)

The Ford Phaeton carried an unusual resuscitator known as the Lung Motor. Simple to operate, oxygen flow could be adjusted according to patient age and size. The Lung Motor worked in much the same way as a modern bicycle floor pump. (Courtesy of NBFD archives.)

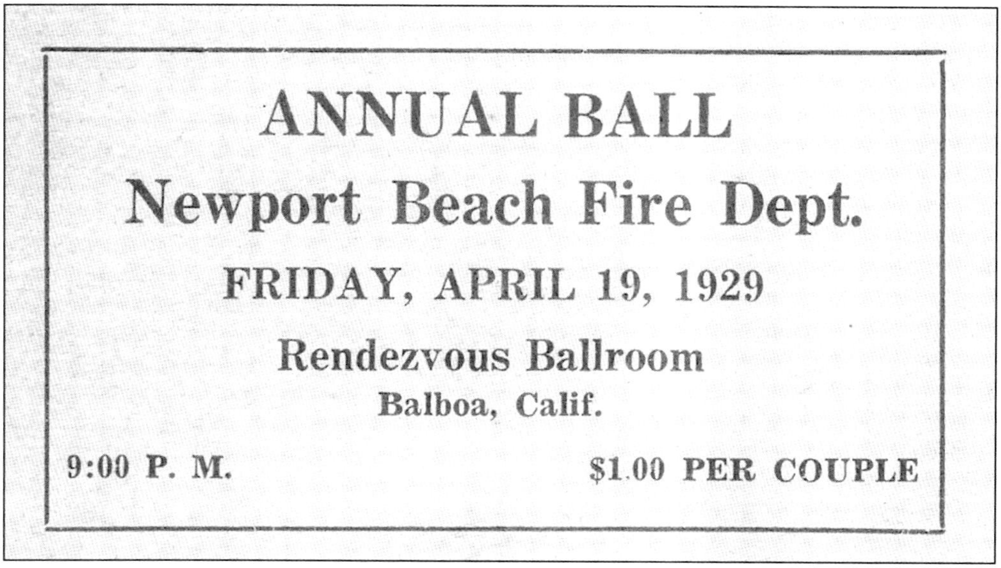
ANNUAL BALL

Newport Beach Fire Dept.

FRIDAY, APRIL 19, 1929

Rendezvous Ballroom

Balboa, Calif.

9:00 P. M. $1.00 PER COUPLE

Social activities continued to be an important aspect of the fire service. Dances and banquets were popular ways for firefighters from neighboring cities to meet, share experiences, and have some fun. This ticket, from 1929, was for a dance held at the famous Rendezvous Ballroom in Balboa, soon to become the site of one of Newport's largest fires. (Courtesy of NBFD archives.)

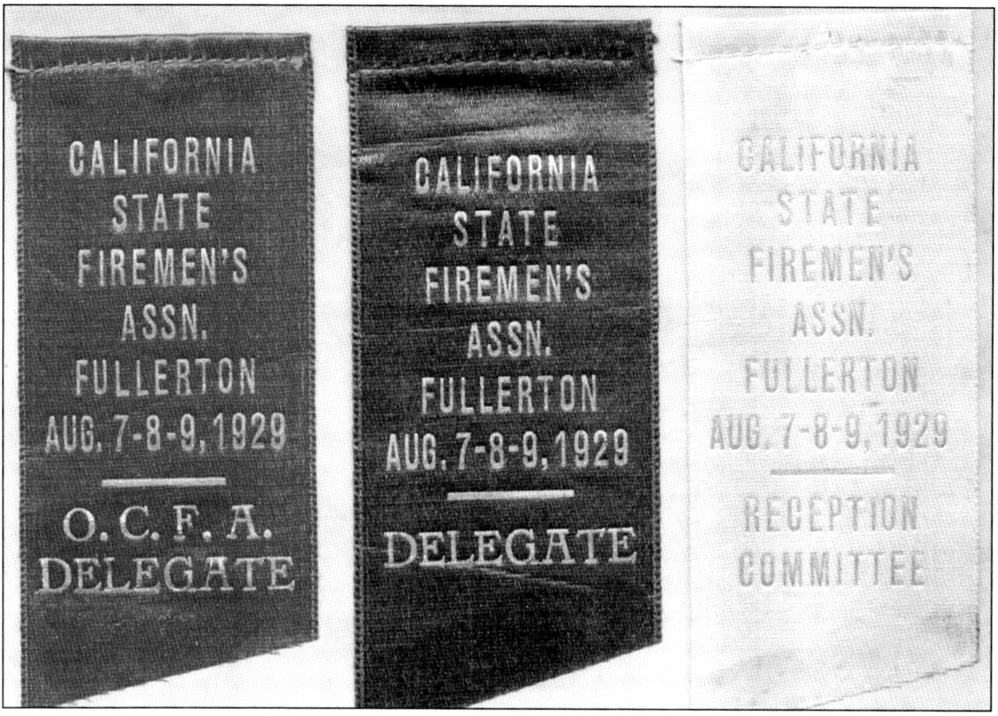

Chief Crocker recognized that membership in organizations like the International Fire Chiefs Association and the newly founded Orange County Fireman's Association was a valuable way to learn the latest in fire-service training and technical innovations. He and Jan Briscoe traveled as far as San Diego and Sacramento to meet with peers and absorb new ideas that could be applied to the Newport Beach Fire Department. (Courtesy of NBFD archives.)

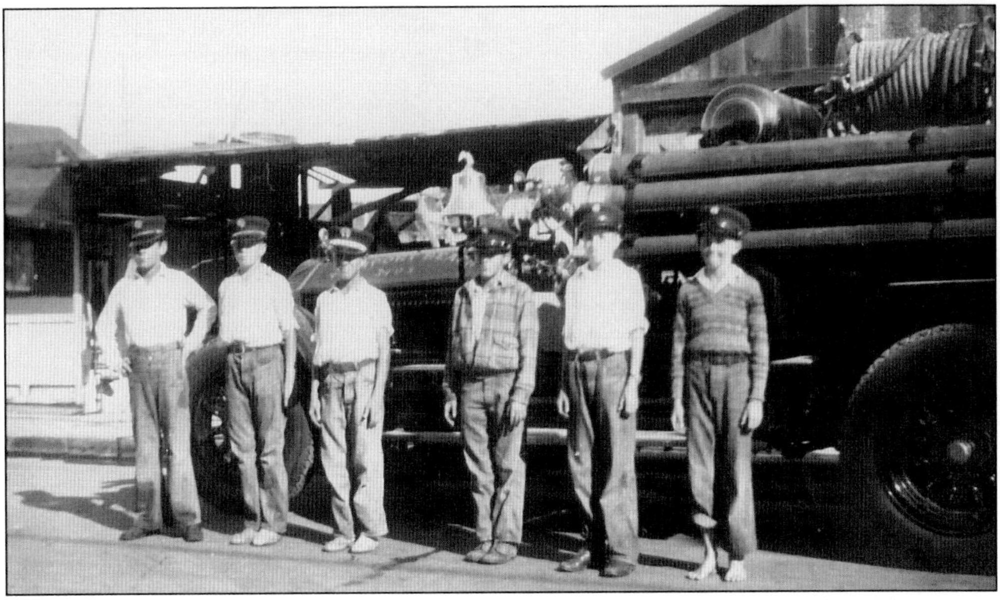

Chief Crocker approached Boy Scout Troop No. 5 of Newport Beach with the idea of forming an apprentice firefighter program. Twenty youths expressed an interest. These Scouts were divided into a hook-and-ladder, hose, and engine company. This program was a precursor to the Explorer programs popular today. (Courtesy of NBFD archives.)

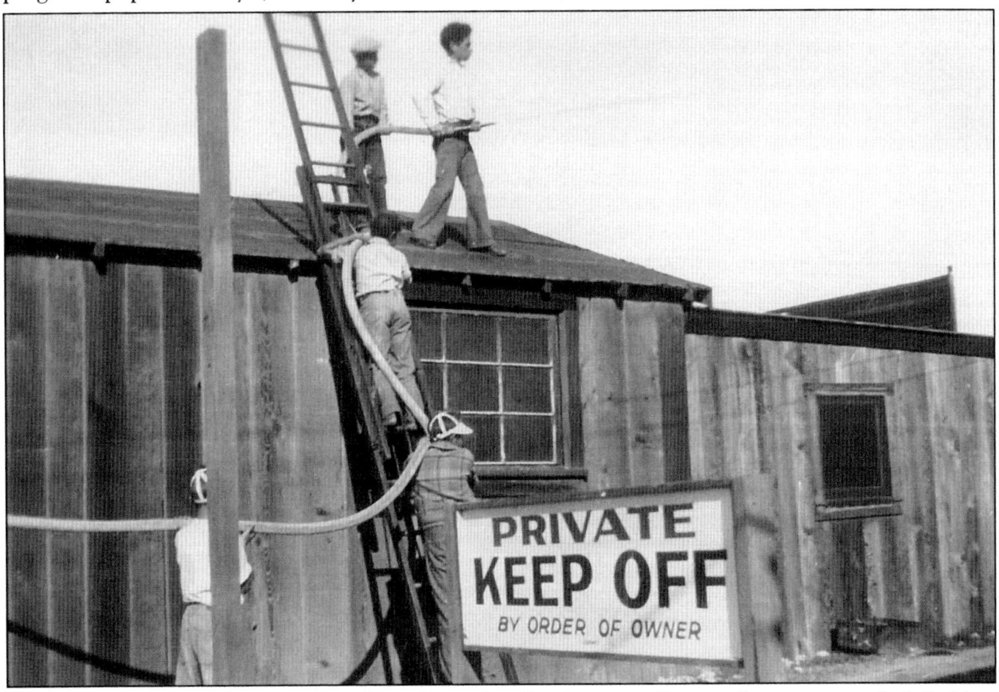

Drills were conducted once a month on Saturday. In this view, Scouts demonstrate their ladder-climbing and firefighting skills. Proficiency tests were given and results determined which Scouts would be appointed as company officers. Merit badges could be earned for demonstrating firefighting aptitude. Several troop members would go on to become professional firefighters with the NBFD. (Courtesy of NBFD archives.)

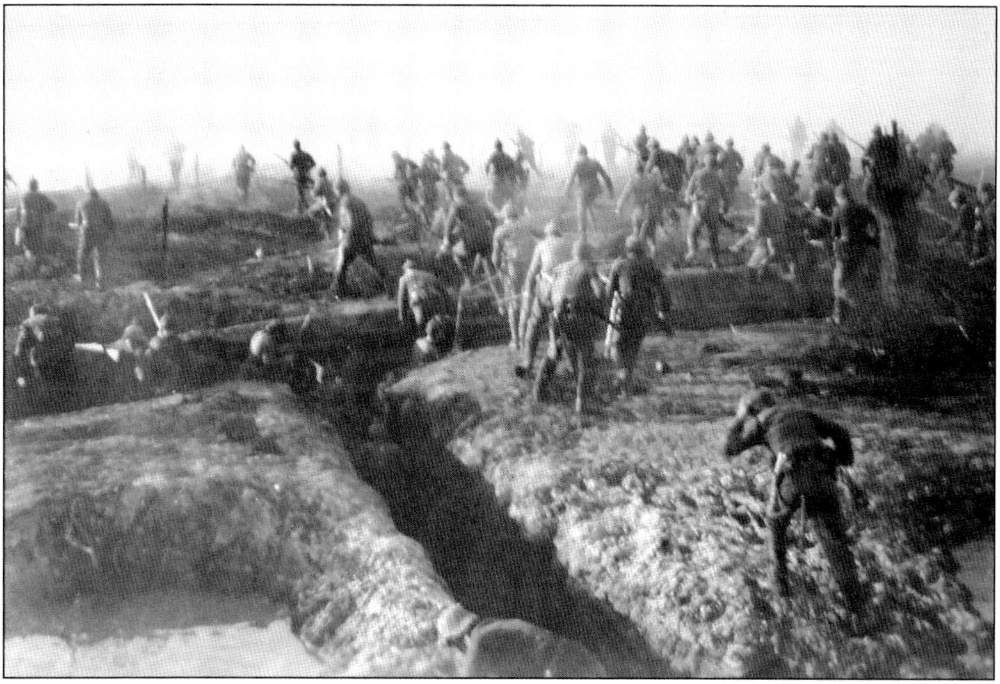

Jan Briscoe was an avid amateur photographer. In 1929, Hollywood came to Newport Beach to film battle scenes for Lewis Milestone's *All Quiet On The Western Front*. Briscoe visited the set with his camera. (Courtesy of NBFD archives.)

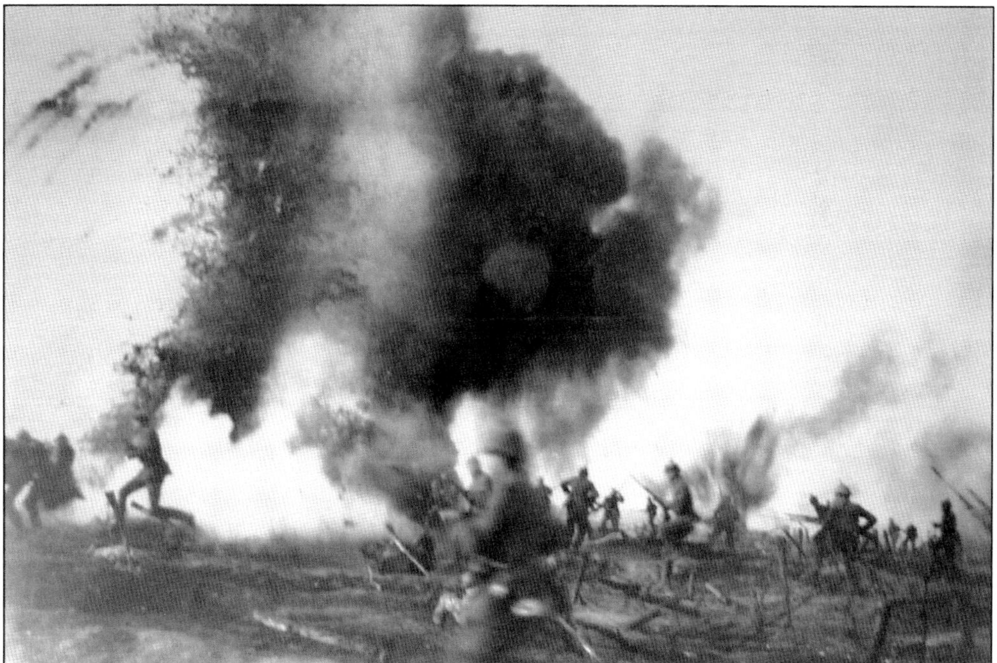

Scenes from *All Quiet On The Western Front* were filmed on the Irvine Ranch south of Corona del Mar. It is likely that Jan Briscoe was there as a fire department representative to insure that the local hills were not set ablaze by special effects pyrotechnics. (Courtesy of NBFD archives.)

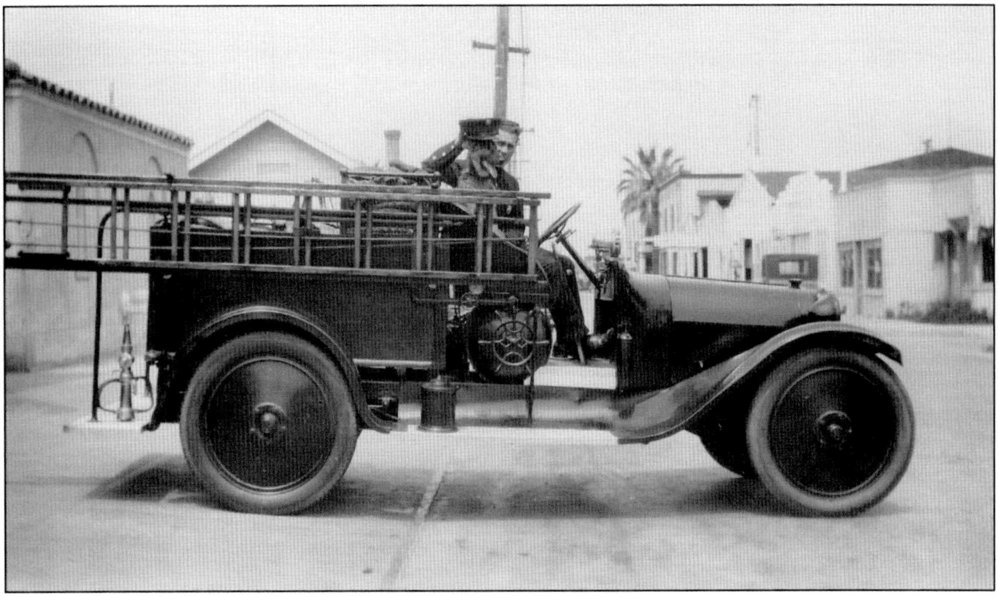

Fireman's Club president Fred Barnett places a uniform cap on Rex, the fire department mascot. In the background, the arcade and amusement-ride complex known today as the Balboa Fun Zone is taking shape. This bay-front location is now the permanent home of the Newport Harbor Nautical Museum. (Courtesy of NBFD archives.)

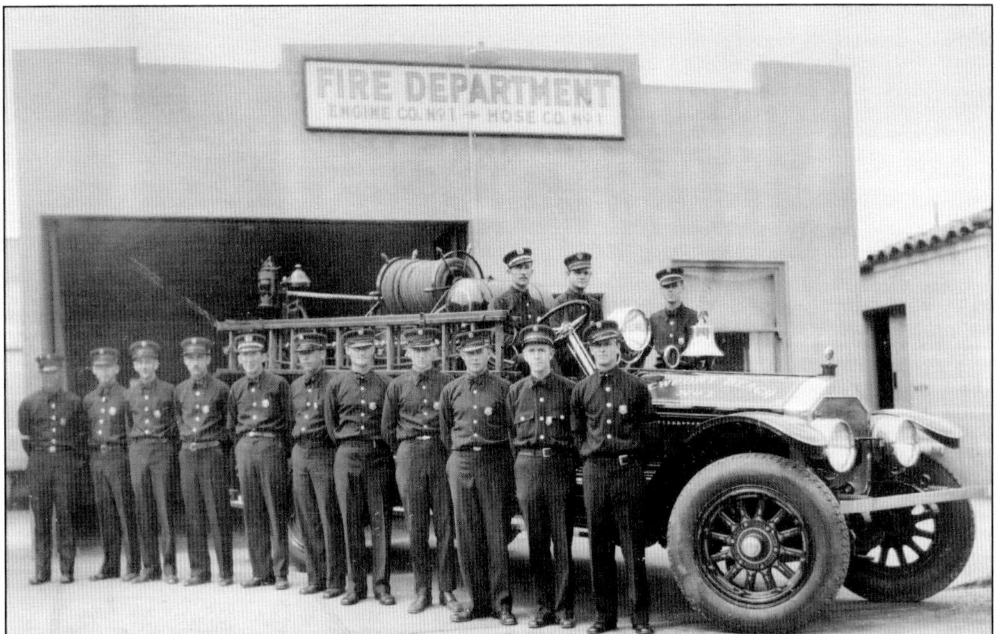

Despite the looming Depression, the city of Newport Beach was on the verge of a healthy transformation. Over the next five years, major dredging and jetty construction would turn mercurial Newport Bay into tranquil Newport Harbor. In turn, the harbor would revitalize and spur residential and commercial development. The fire department would grow along with the city. (Courtesy of NBFD archives.)

Three

Balboa Island

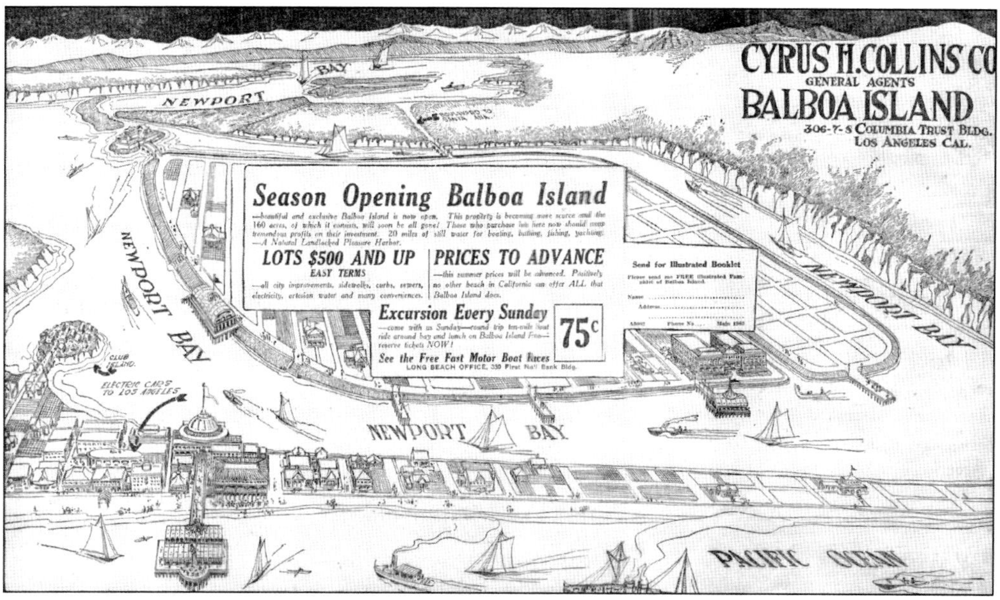

Balboa Island sits in Newport Harbor opposite the Balboa Pavilion. Around 1908, real estate developer William Collins used dredged sand to give permanence to what was once a shifting sandbar. Summer cottages began springing up. In 1916, the island was incorporated into the city of Newport Beach. Visitors to Balboa could take a ferry to the island and stroll along its seawall. More and more people began calling Balboa Island their home. In 1929, the wood bridge to the island was replaced with a concrete span. Cars, and fire engines, could now travel on and off the island with ease. (Courtesy of NBFD archives.)

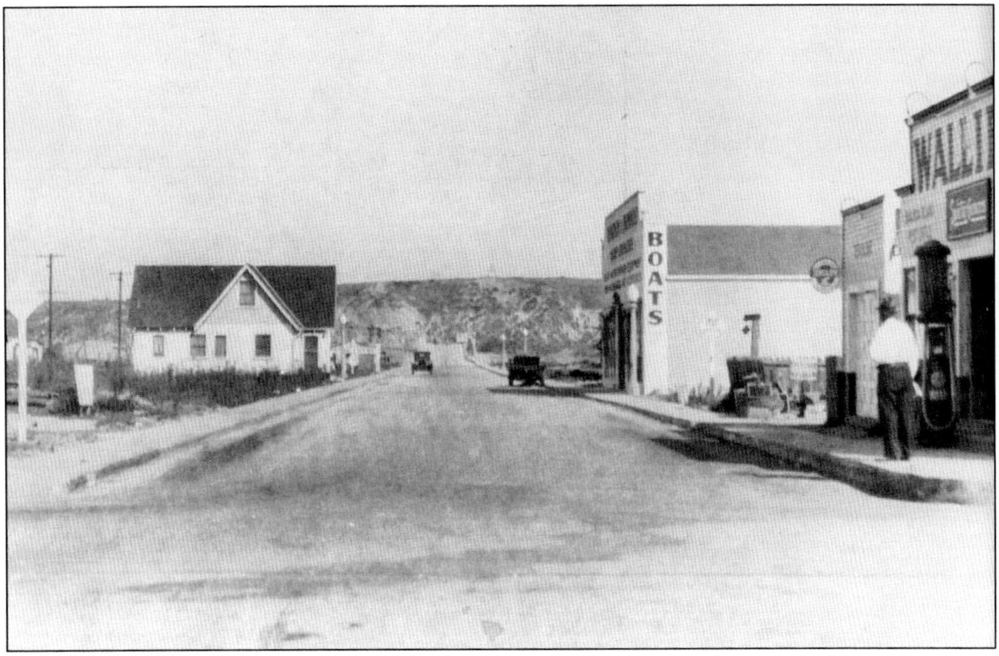

By 1927, Balboa Island residents were demanding a piece of the fire-protection pie. The Dodge Squad was moved onto the island in July after Station One took delivery of the new American LaFrance pumper. Marine Avenue, as this photograph shows, was just developing into the island's commercial center. The Squad was temporarily housed in the Boat Garage, visible on the right. (Courtesy of NBFD archives.)

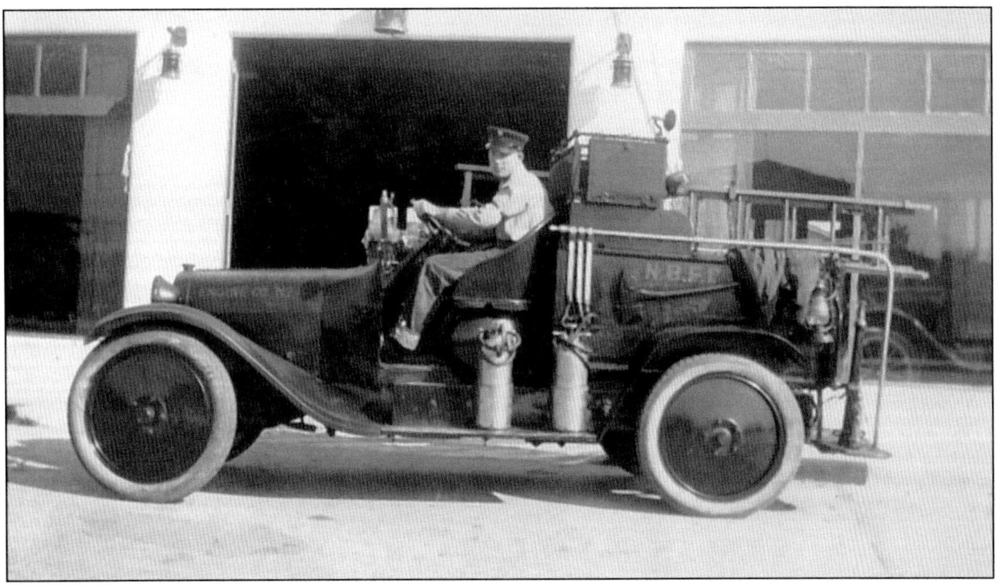

The Dodge Squad served the city of Newport Beach for 30 years, morphing along the way from the ancient Rambler No. 2 into a versatile utility vehicle. Between 1921 and 1927, it was the only motorized fire apparatus in Newport Beach. In this photograph, the Squad sits in front of the Boat Garage on Marine Avenue, now designated as Newport Engine No. 4. This engine was loaded with all the equipment necessary to handle most emergencies. (Courtesy of NBFD archives.)

Construction on Balboa Island's fire station started in October 1930. A small lot was purchased on Marine Avenue and leveled using a horse and weighted plow. Volunteer firefighter Andy Gengler holds the reins. (Courtesy of NBFD archives.)

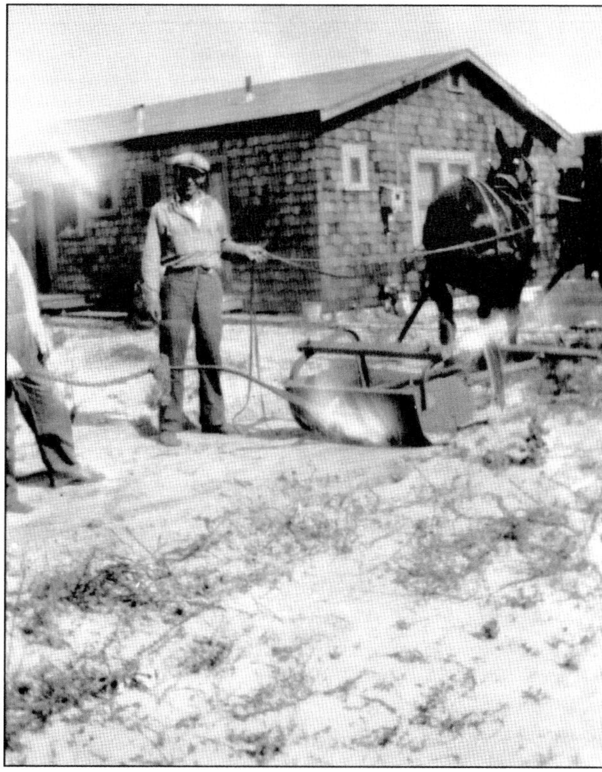

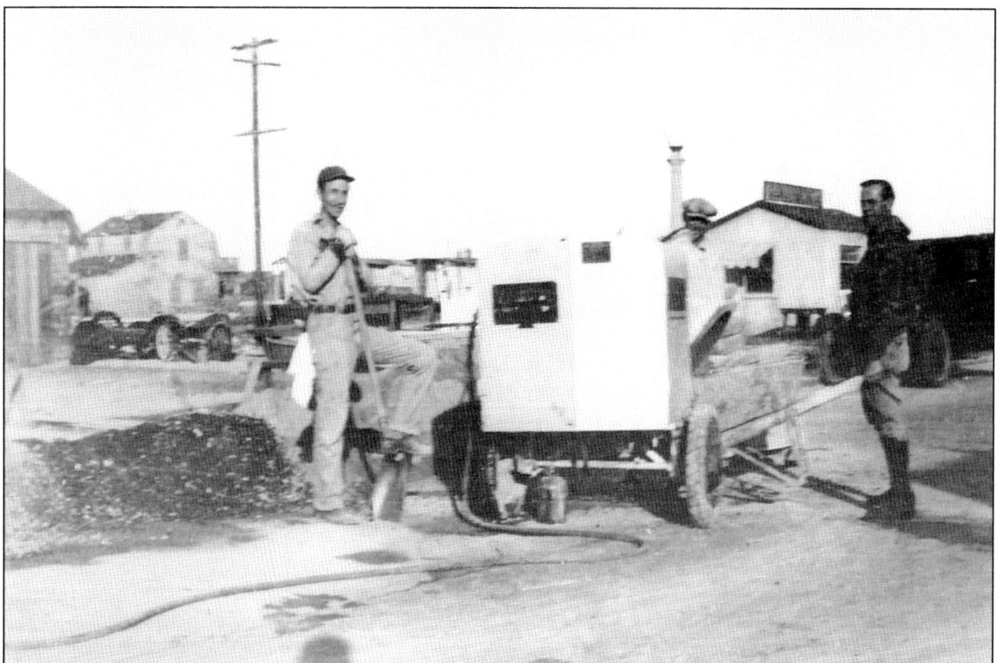

Several paid-call firefighters, including Andy Gengler, were utilized to build the new Balboa Island fire station, soon to be designated as Station Four. Gengler later served as a paid-call firefighter on Balboa Island. (Courtesy of NBFD archives.)

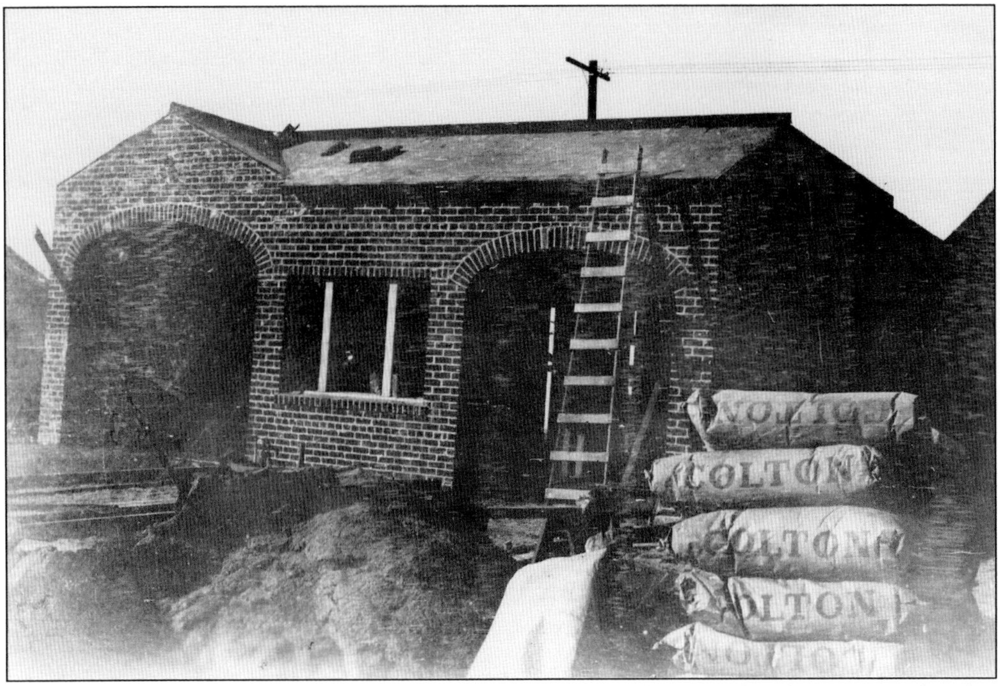

The new station was a wood-frame building with a brick facade covered with stucco. Today this unusual facade is protected by historical landmark status. (Courtesy of NBFD archives.)

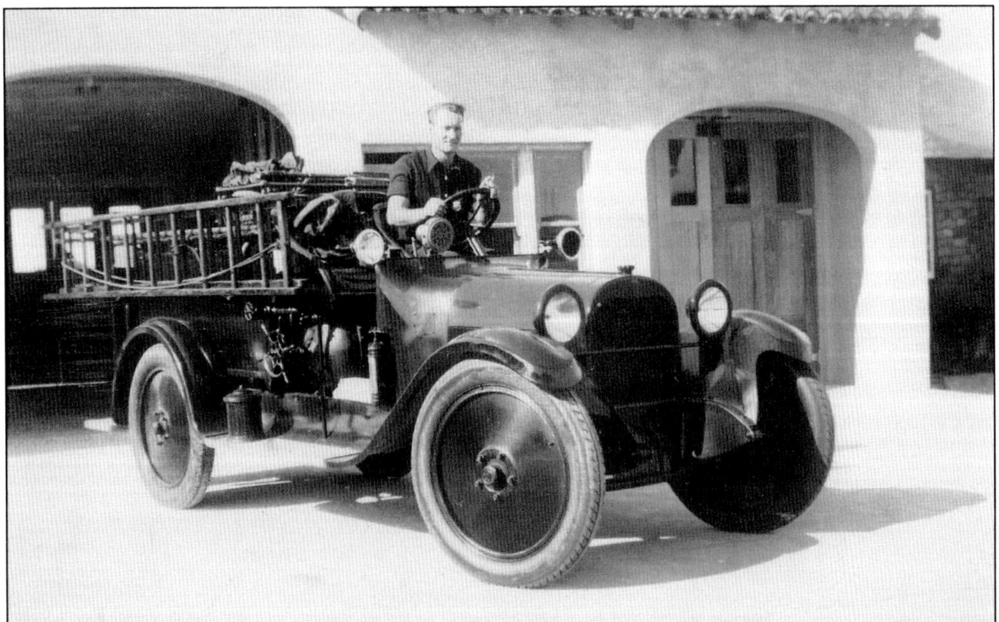

Andy Gengler sits at the wheel of the Dodge Squad. It appears that Andy was given the honor of driving the Squad over from the Boat Garage and backing it into its new station. Station Four opened on January 1, 1931. (Courtesy of NBFD archives.)

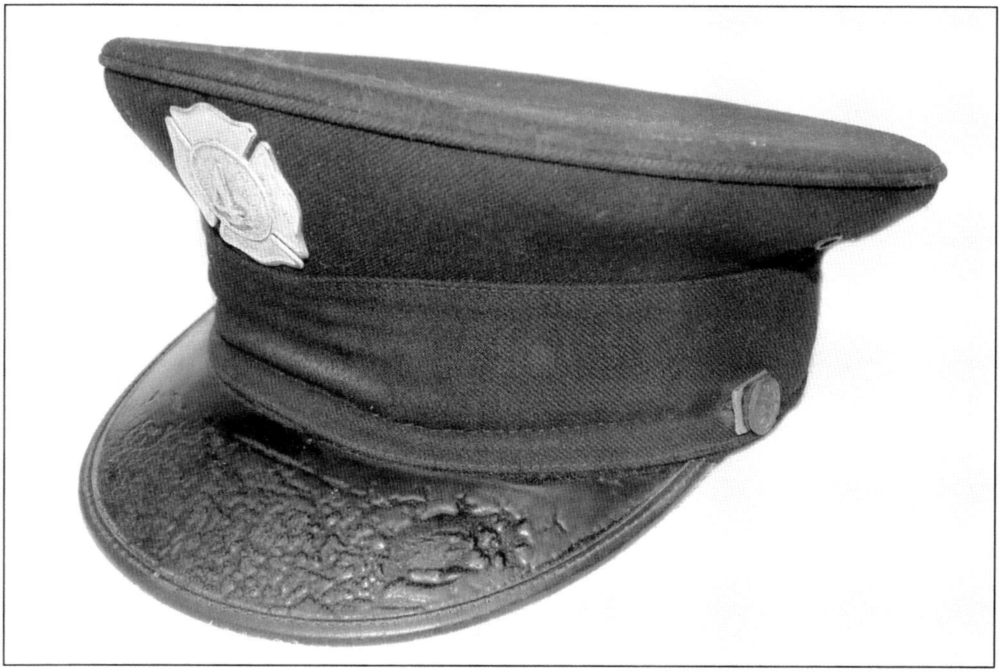

During the 1940s, Newport Harbor High School students were recruited and trained to respond to calls as volunteer firefighters. Capt. Ed Zube would pull Engine No. 4 out of the station when the alarm sounded, wait a few seconds for these youngsters to arrive, or meet them at the scene. Dr. Alan Andrews, a lifelong Newport Beach resident, was one of those young men. This was his fire cap. (Courtesy of NBFD archives; gift of Mrs. Betty Andrews.)

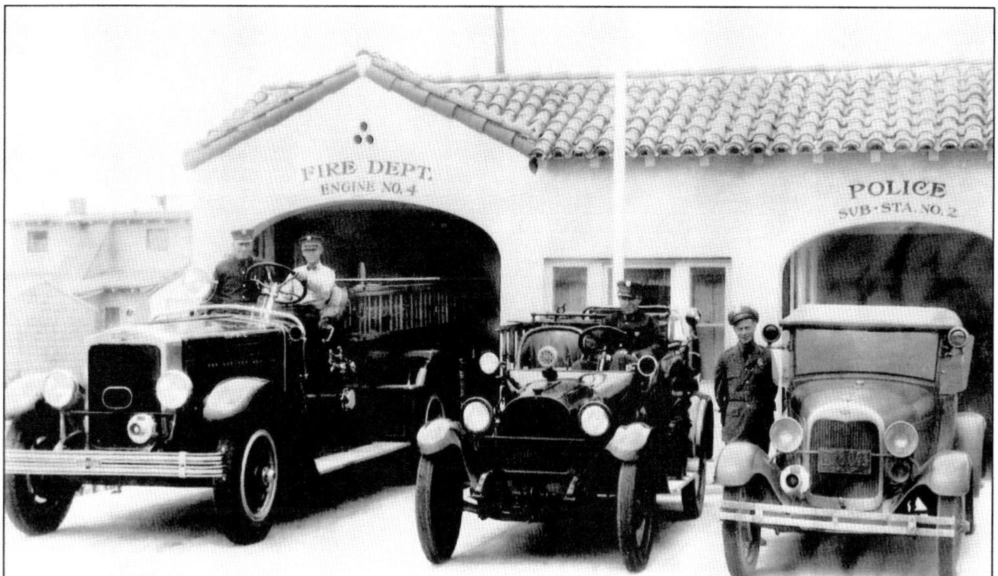

In subsequent years, Station Four was shared with the Newport Beach Police. As fire engines grew larger, this station seemingly got smaller. The police eventually moved elsewhere. Before the station was closed in 1994, it housed a three-man fire crew in its quaint but cramped quarters. (Courtesy of NBFD archives.)

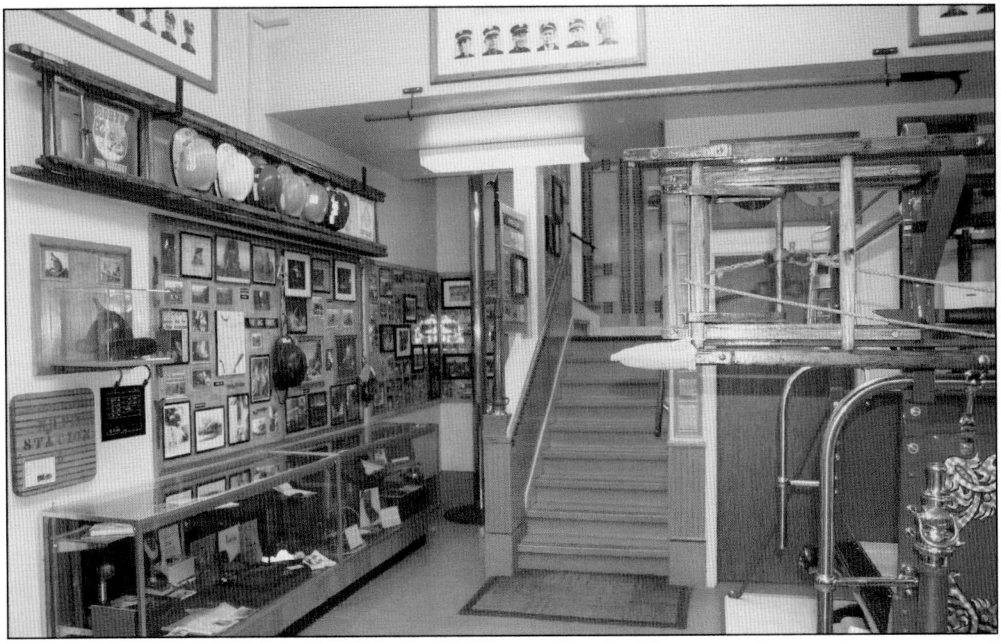

In 1994, the original Balboa Island Station Four was closed. A large, modern station was erected at the opposite end of Marine Avenue. Today, in addition to housing a fire company that has proudly served the residents of Balboa Island and the city of Newport Beach for over 75 years, Station Four is the home of Newport Beach Fire Department's restored 1920 American LaFrance and an impressive display of historical memorabilia. (Author's collection.)

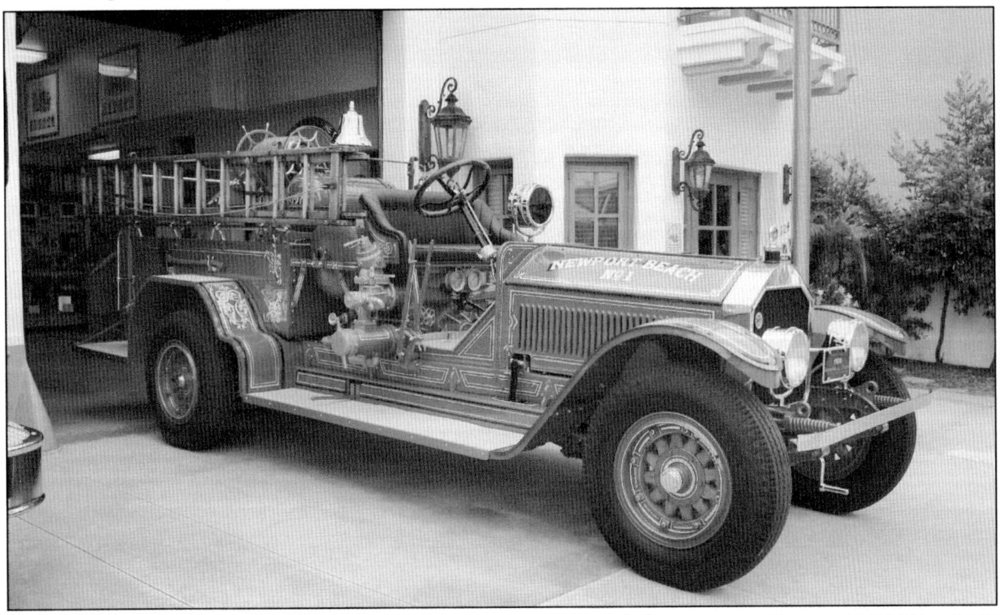

Sadly, none of Newport Beach's original fire engines survived the passage of time. In 2000, a 1920 American LaFrance engine was donated to the city. A three-year restoration of this engine was completed in 2006. Rick Poulsen, a longtime city employee and automobile restorer extraordinaire, did the work. It remains one of the finest American LaFrance restorations in the country today. (Author's collection.)

Four

THE GROWTH YEARS

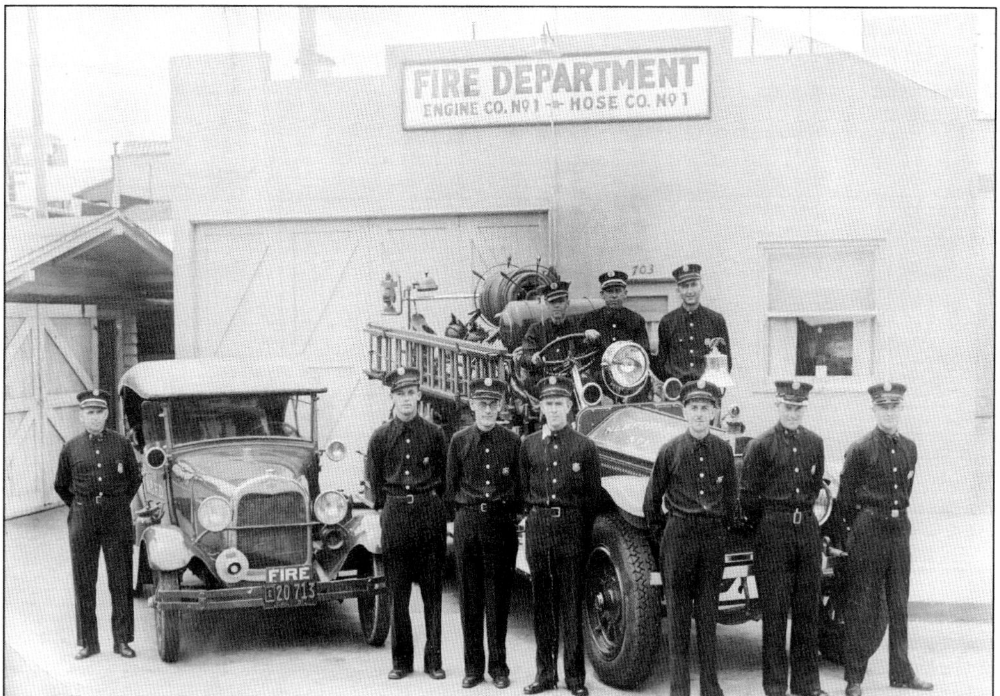

A pivotal year for the Newport Beach Fire Department was 1931. Two additional fire stations were built, two new fire engines were purchased, and three men were added to the payroll. Things would remain stable until the postwar years, when prosperity and optimism fueled the rapid growth in both the city of Newport Beach and its fire department that continues to this day. Five additional fire stations, a paramedic program, an expanded fire prevention bureau, and other programs would require a workforce of nearly 150 men and women by the year 2000. (Courtesy of NBFD archives.)

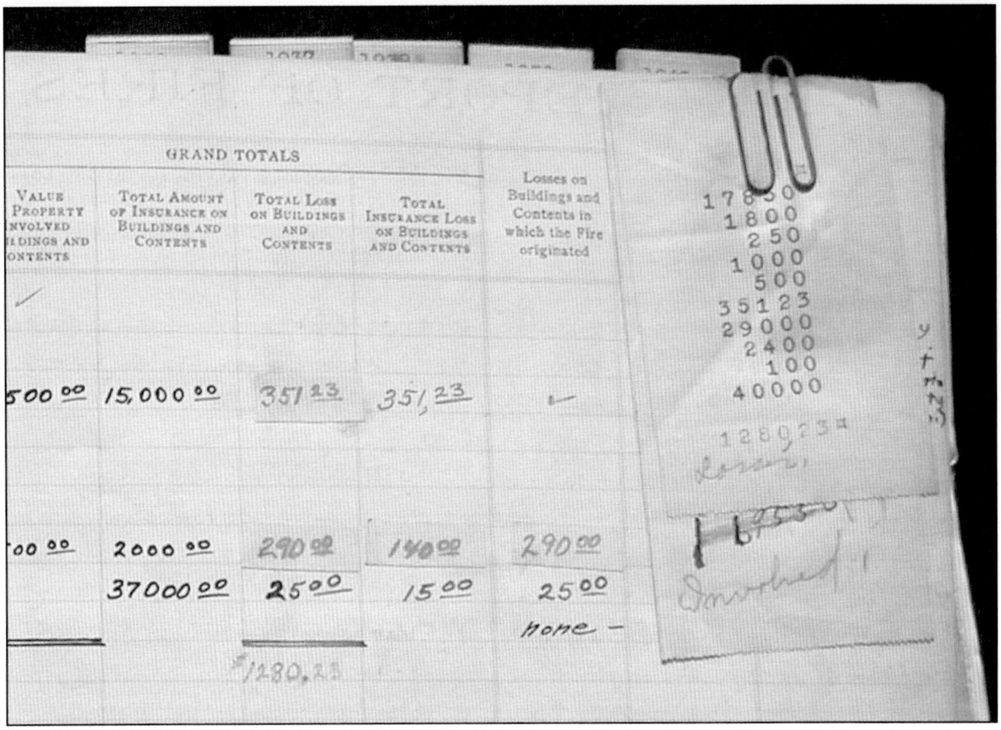

Chief Crocker's record keeping was legendary. At the end of each year, he pulled out his adding machine and totaled the fire losses. These per-capita losses were important, as they determined insurance rates. For the most part, each year's total losses were very low, a source of pride for the chief and the fire department. (Courtesy of NBFD archives.)

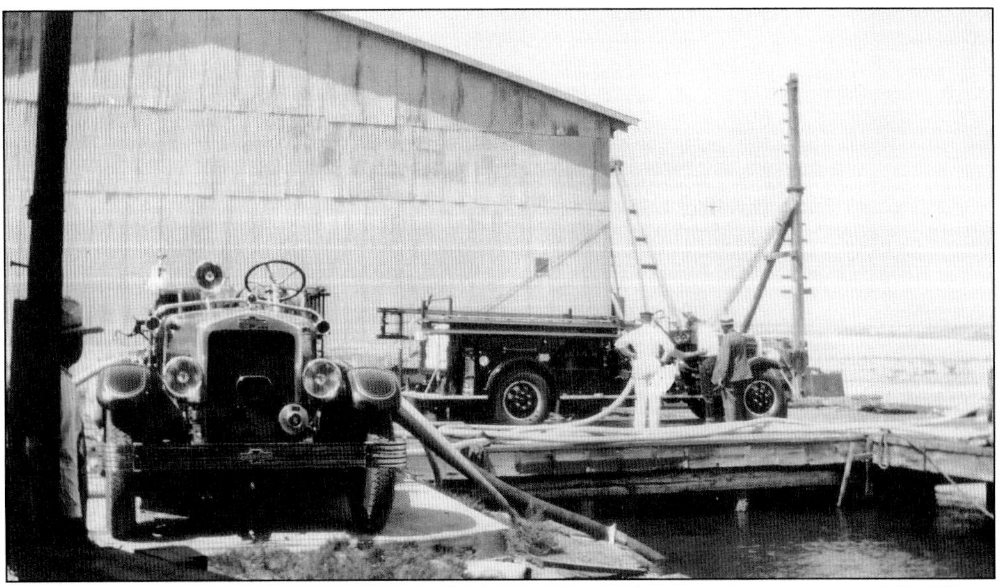

The NBFD tripled its fire engine fleet with the purchase of two new pumpers in 1931. These engines, undergoing their acceptance test in this view, would be housed in two new fire stations being built that same year. (Courtesy of NBFD archives.)

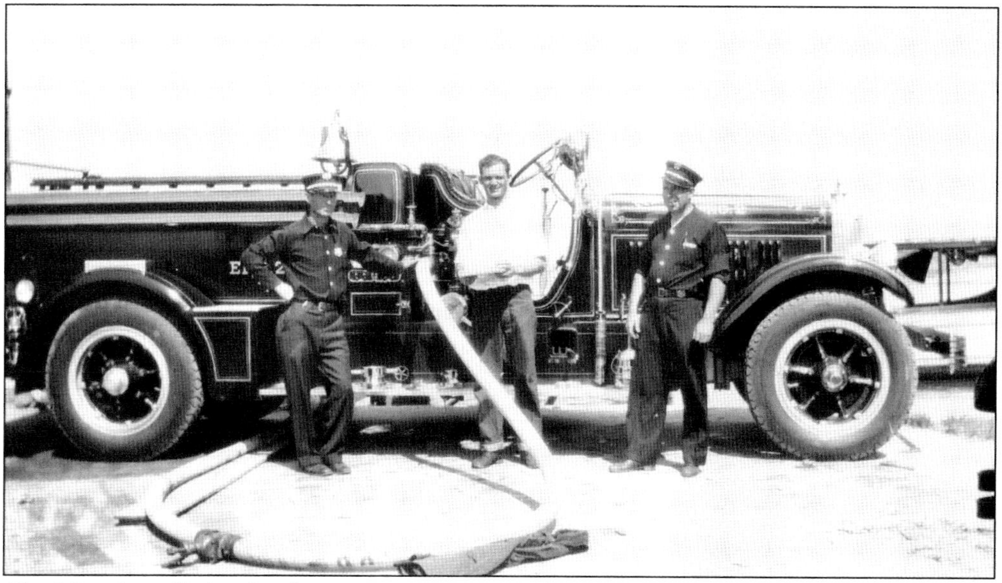

Chief Crocker rolled up his sleeves to assist Capt. Walt Honeycutt with Engine No. 2's acceptance test. No. 2 was an American LaFrance Type 12 triple-combination fire engine. It cost $14,605. (Courtesy of NBFD archives.)

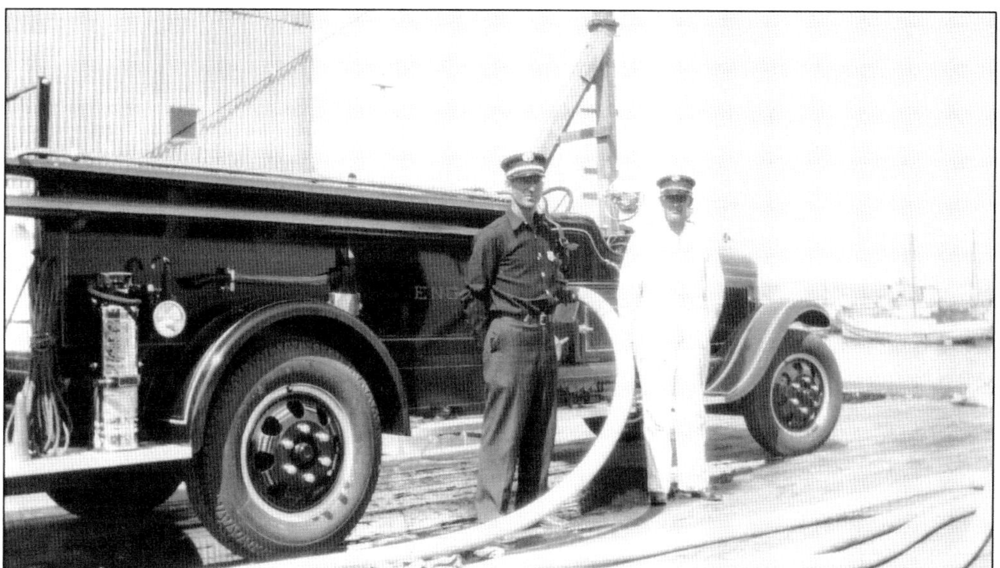

Capt. Howard Roberds, in white, stands next to Newport Engine No. 4, a GMC/American LaFrance Type 99 triple-combination fire engine. No. 4 cost $6,304. This engine was destined for duty on Balboa Island. Howard Roberds was appointed captain at Station Four. (Courtesy of NBFD archives.)

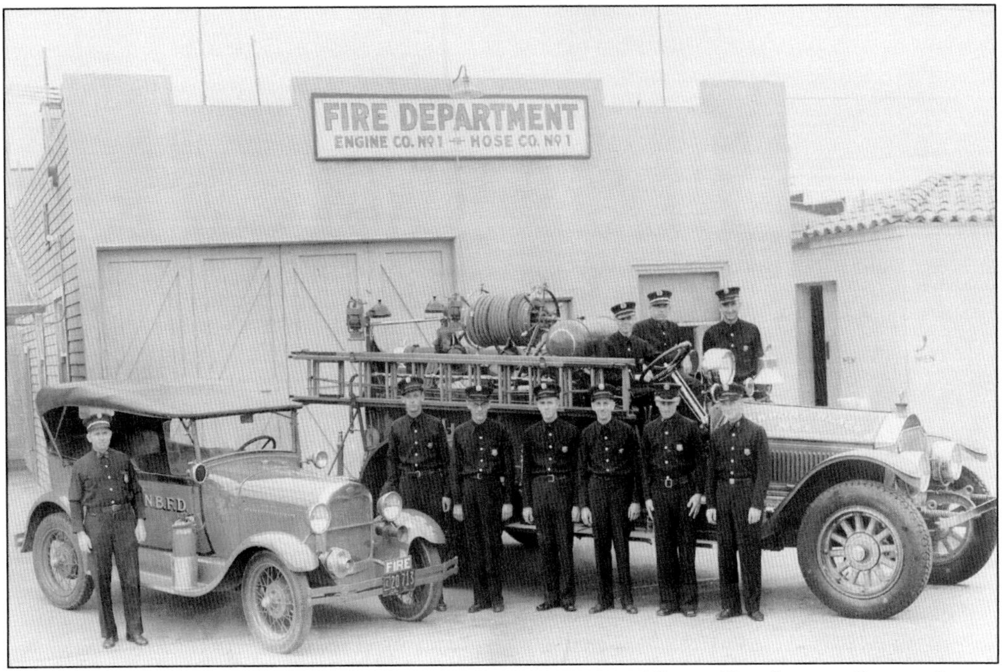

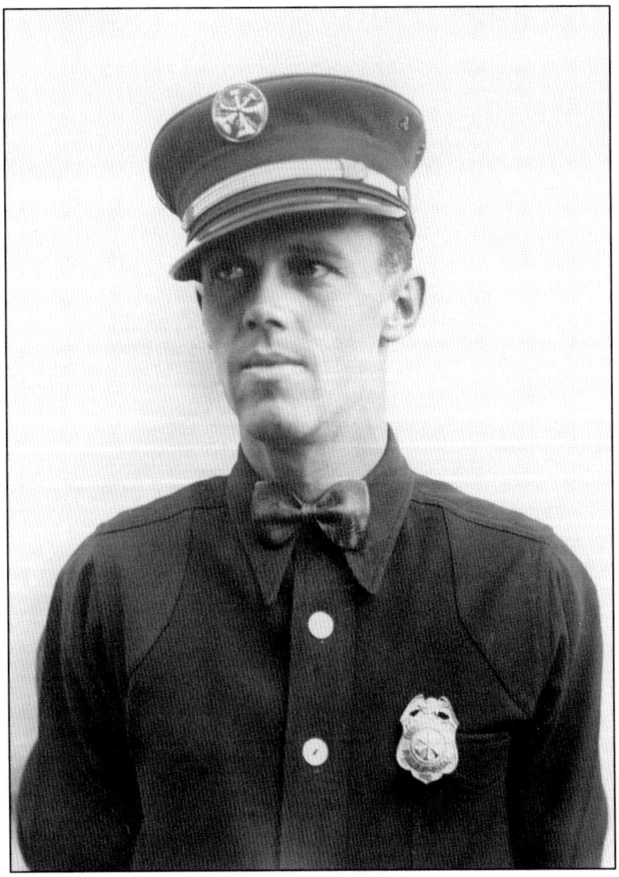

This is Fire Station One and crew as they appeared in 1931. Chief Crocker stands proudly next to his Model A Phaeton. A paid firefighter was assigned to each station. Each worked a 12-hour shift, which started around 7:00 a.m. and ended at 7:00 p.m. A relief driver was assigned to each station, as were eight paid-call firefighters. Sleeping accommodations were furnished free to paid-call members. (Courtesy of NBFD archives.)

Chief Crocker appointed Jan Briscoe as fire captain of Station One. Briscoe was the third paid firefighter hired by the Newport Beach Fire Department. (Courtesy of Don Crocker.)

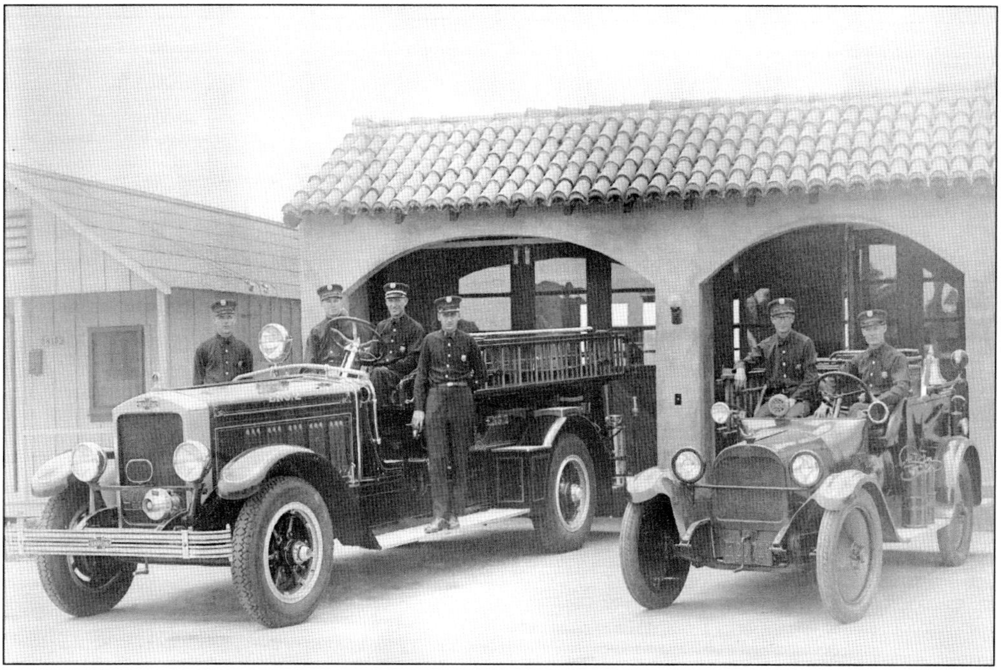

Station Two was opened in June 1931, at 2817 West Central Avenue, alongside the nearby Pacific Electric Red Car line. Station Two was unique at the time. It was Newport's first two-bay station, with room for two engines. Twenty-one years after the first hose reel shack was built, the Newport Pier area finally had its own fire station. (Courtesy of NBFD archives.)

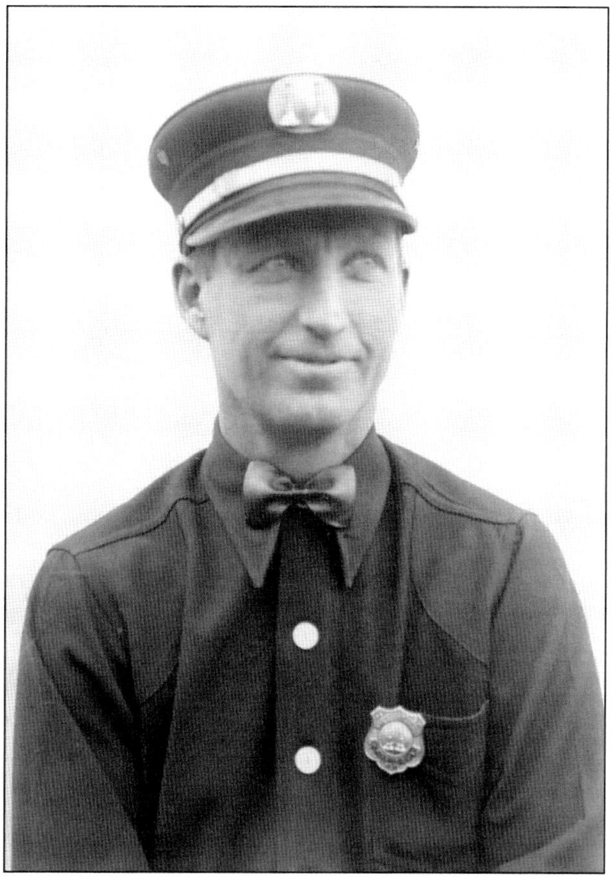

Walt Honeycutt was the fourth firefighter hired as a full-time employee. He was appointed captain at Station Two. One night, in gale force winds and pelting rain, an unconscious employee of the Corona del Mar bathhouse was pulled from the bay. Firefighters worked at resuscitating this drowning victim for seven hours, despite his being declared dead by a doctor. Their perseverance was rewarded when a faint breath was finally detected. Walt Honeycutt was one of those rescuers. (Courtesy of Don Crocker.)

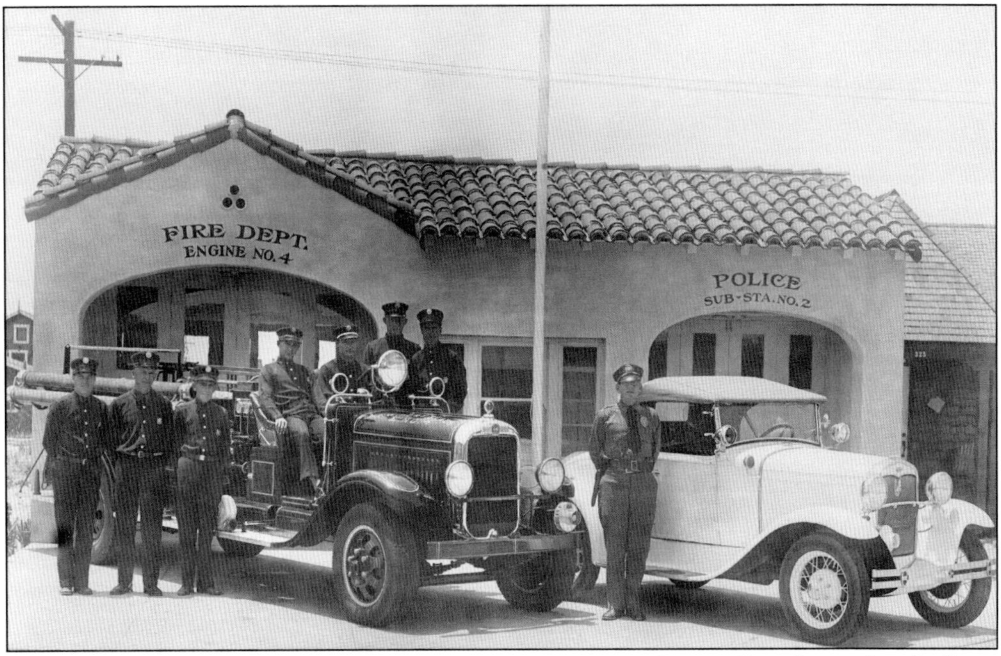

Station Four, located on Balboa Island, was opened in January 1931. Bridges over the Santa Ana River and Newport's upper bay were in place by 1930, making travel along Pacific Coast Highway easier. The east side of Newport Beach was now more accessible and subject to increased development. Station Four was situated to provide fire protection to Balboa Island and Corona del Mar. Howard Roberds sits in the captain's seat. (Courtesy of NBFD archives.)

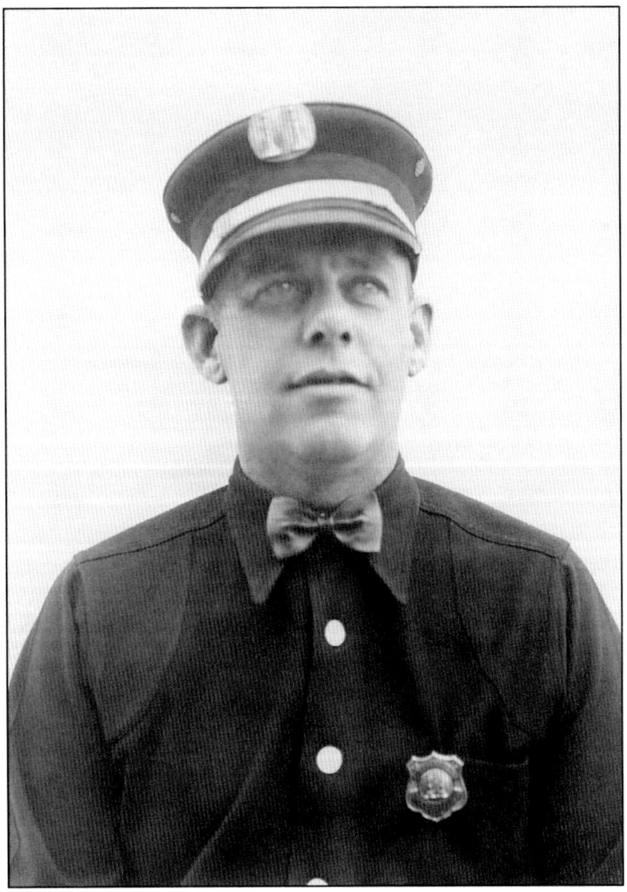

Clinton Dickey left the fire department in 1930. Perce Pellett was promoted from paid-call firefighter to full-time assistant chief. The department now had a complement of five professionals along with a reliable force of paid-call firefighters who could respond to calls and work night shifts in the stations. (Courtesy of Don Crocker.)

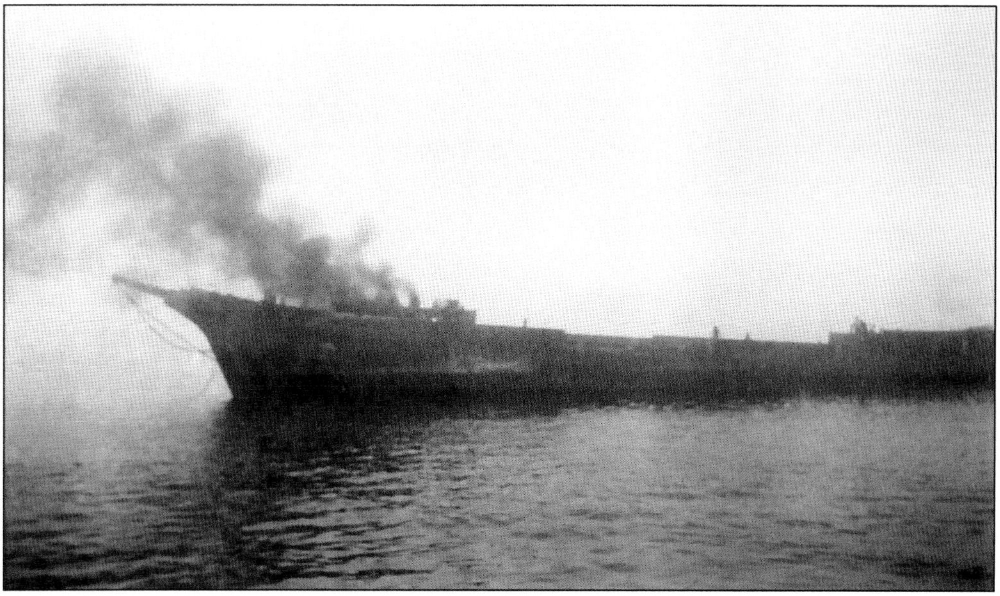

Remember the *Muriel*, the schooner that volunteer fire chief Rube Shafer attempted to tow out to sea in 1925? After five years of hindering navigation near the harbor entrance, and serving as a home for at least one very large octopus, it burned to the waterline. Dynamite was used to finish the job. (Courtesy of NBFD archives.)

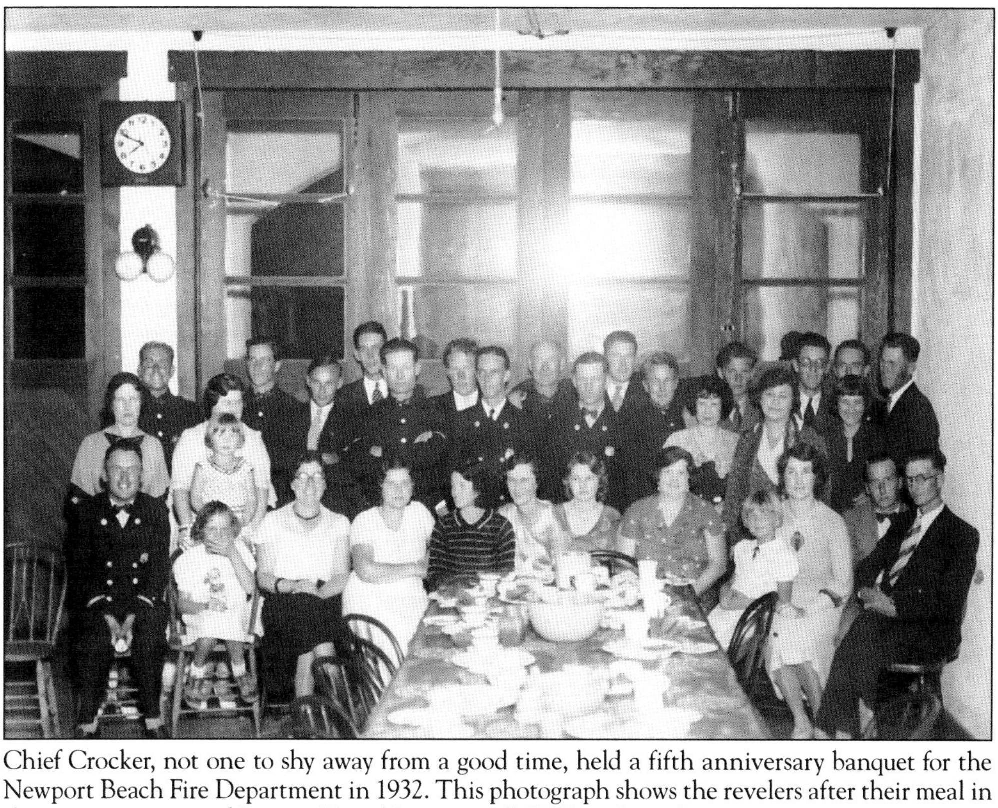

Chief Crocker, not one to shy away from a good time, held a fifth anniversary banquet for the Newport Beach Fire Department in 1932. This photograph shows the revelers after their meal in the apparatus room of Station Two. (Courtesy of NBFD archives.)

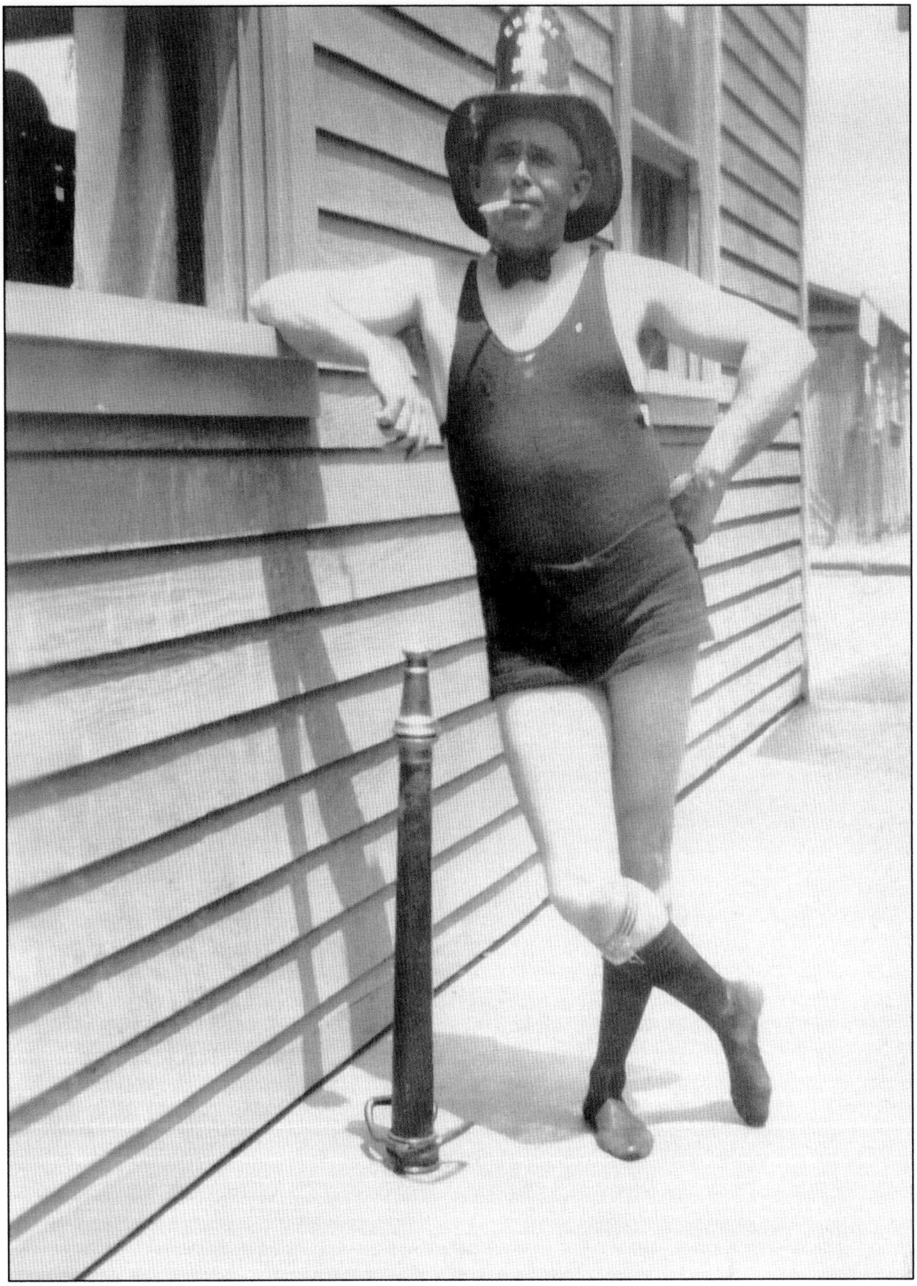

In this photograph, Chief Crocker comically poses in his dual role as fire chief and head of the lifeguard service. Prior to 1927, lifesaving efforts in Newport Harbor and along the beaches were unorganized and undermanned. High surf and shifting sandbars made the harbor entrance particularly dangerous to those unfamiliar with its waters. Several boating accidents claimed multiple victims. In one instance, Newport Beach volunteer lifeguard and Hawaiian surfing legend Duke Kahanamoku paddled out and saved several lives. Kahanamoku had founded a surfing club in Corona del Mar and lived in a tent on one of its beaches. Despite the heroics of Kahanamoku and a few others, it was becoming evident that a professional, organized approach to water rescue and safety was necessary. (Courtesy of NBFD archives.)

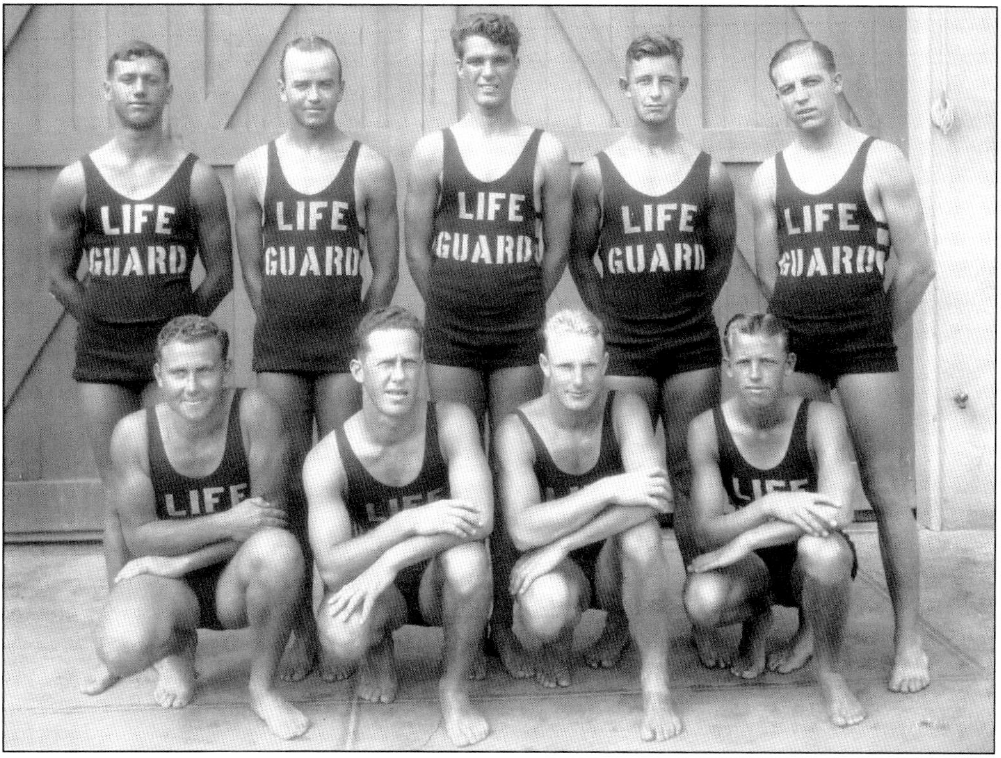

In 1928, Chief Crocker established a lifeguard service consisting of 10 full-time men officially affiliated with the fire department. One of the more interesting responsibilities of the lifeguard corps was to enforce Newport's indecency laws. Lifeguards routinely patrolled the beach, keeping their eyes peeled for bathers in scanty attire. Daredevils wearing trunks or rolled-down suits were subject to arrest. (Courtesy of NBFD archives.)

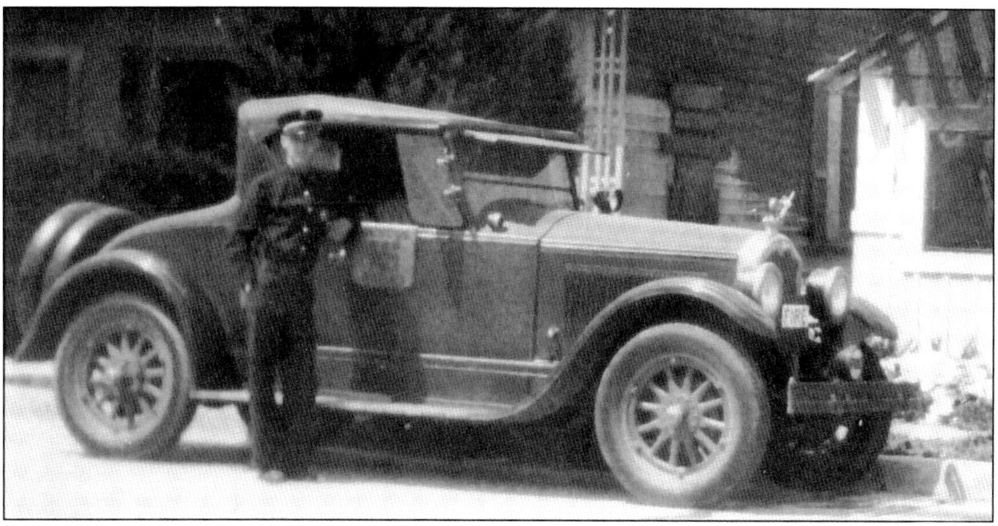

A 1926 Buick Roadster was used for both lifesaving and fire-related emergencies. A sign on the door identified it as the NBFD Rescue Squad. The Buick's mobility and equipment, along with lifeguards on the beach, combined to provide city residents and visitors with an effective lifesaving service. (Courtesy of NBFD archives.)

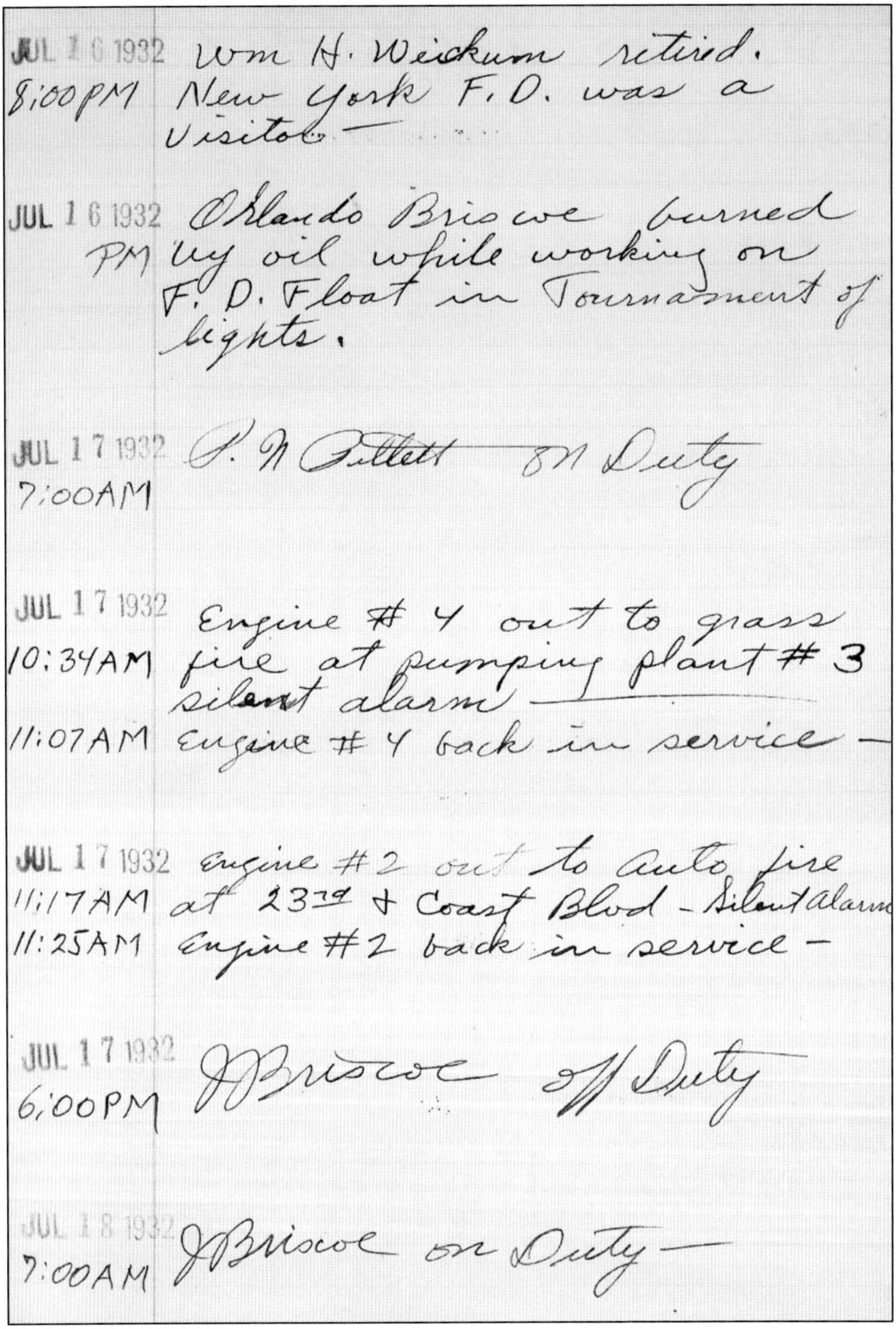

JUL 16 1932 8:00PM Wm H. Weickum retired. New York F.D. was a visitor —

JUL 16 1932 PM Orlando Briscoe burned by oil while working on F. D. Float in Tournament of lights.

JUL 17 1932 7:00AM P. N. Pittett on Duty

JUL 17 1932 10:34AM Engine # 4 out to grass fire at pumping plant # 3 silent alarm —
11:07AM Engine # 4 back in service —

JUL 17 1932 11:17AM Engine #2 out to auto fire at 23rd & Coast Blvd - Silent Alarm
11:25AM Engine #2 back in service —

JUL 17 1932 6:00PM J Briscoe off Duty

JUL 18 1932 7:00AM J Briscoe on Duty —

Newport Beach engine companies began keeping daily logbooks in 1932. Most entries documented routine personnel changes and apparatus maintenance. All emergency calls were logged. There were some fascinating entries, like the 1933 Long Beach Earthquake and the mandated blackouts that followed December 7, 1941. Today these journals, filled with over 60 years of department history, reside in the department's archives. (Courtesy of NBFD archives.)

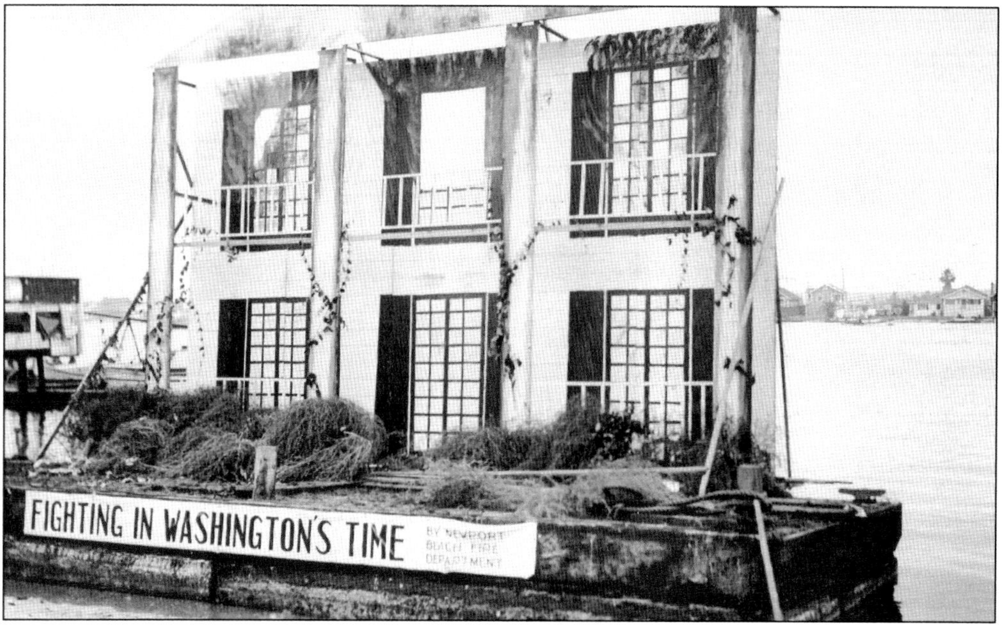

In 1907, Newport Bay gondolier John Scarpa attached several Japanese lanterns to his gondola. In 1908, a lighted boat parade, inspired by Scarpa, was inaugurated and a tradition was born. This parade soon became the annual Tournament of Lights, complete with floats inspired by the Pasadena Rose Parade. In 1932, the Newport Beach Fire Department entered a float of its own. Its theme was "Fire Fighting In [George] Washington's Time." (Courtesy of NBFD archives.)

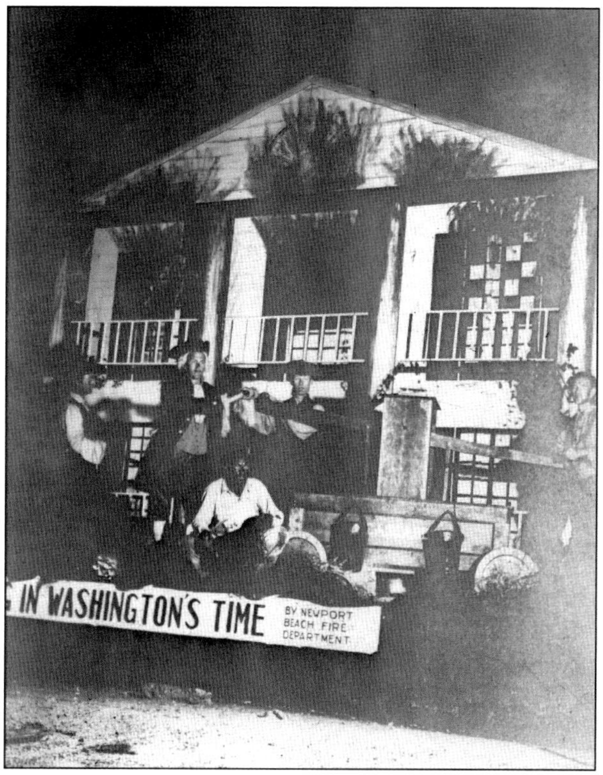

Newport Fire's float featured a two-story mansion complete with simulated fire and firefighters dressed in colonial costume. The centerpiece was a working reproduction of a 1792 hand pumper called the Cascade. Newport Fire took first place in its division. (Courtesy of NBFD archives.)

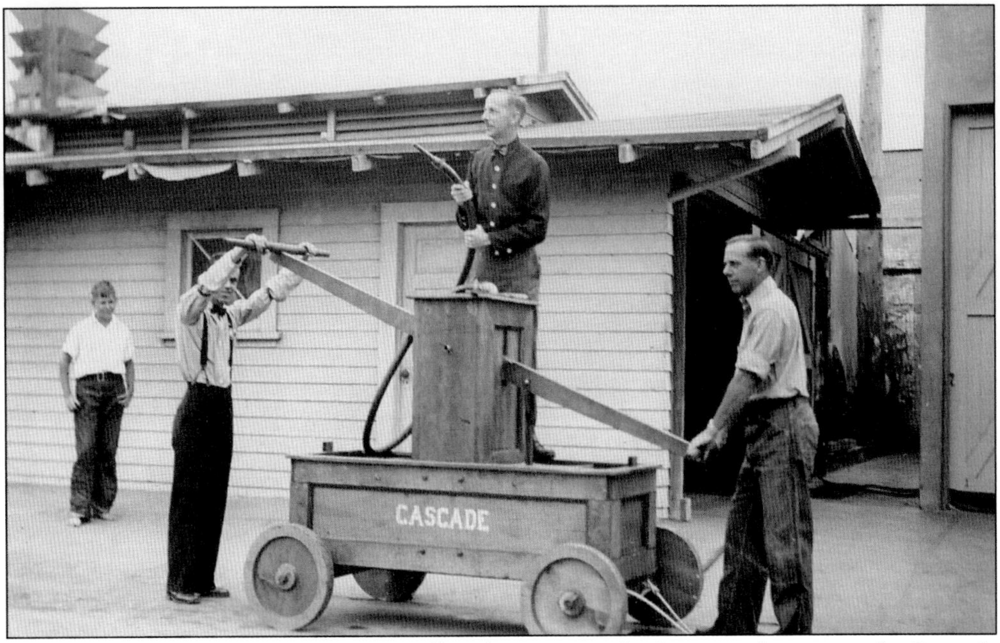

Jan Briscoe and brothers Lee and Perce Pellett demonstrate the finer art of operating the Cascade hand pumper. Lifeguard headquarters, adjacent to Station One, provides the backdrop. This hand pumper could shoot a stream of water 60 feet. The Cascade remains on display 75 years later at Newport's largest fire station. (Courtesy of NBFD archives.)

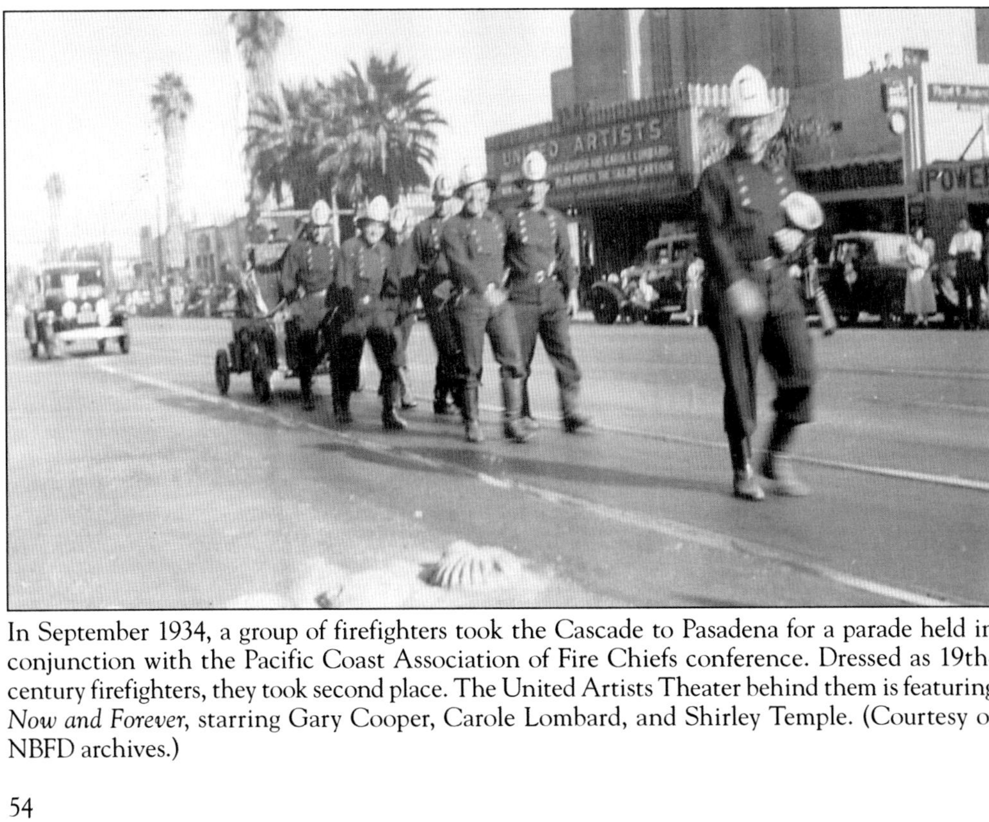

In September 1934, a group of firefighters took the Cascade to Pasadena for a parade held in conjunction with the Pacific Coast Association of Fire Chiefs conference. Dressed as 19th-century firefighters, they took second place. The United Artists Theater behind them is featuring *Now and Forever*, starring Gary Cooper, Carole Lombard, and Shirley Temple. (Courtesy of NBFD archives.)

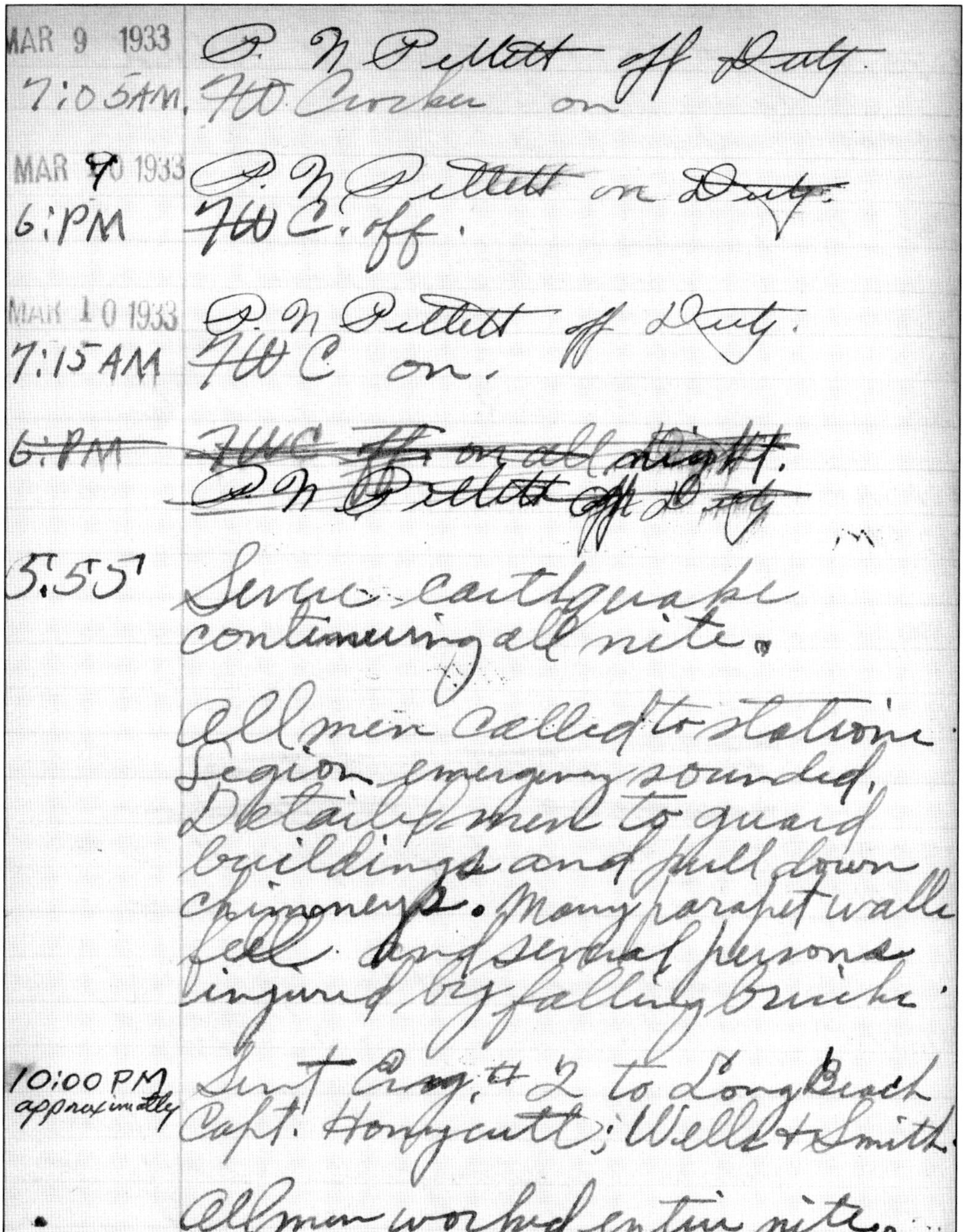

MAR 9 1933 P. N. Pellett off Duty
7:05 AM. F.W. Crocker on

MAR 9 10 1933 P. N. Pellett on Duty.
6:PM F.W.C. off.

MAR 10 1933 P. N. Pellett off Duty.
7:15 AM F.W.C on.

~~6:PM [illegible]~~
~~P. N. Pellett off Duty~~

5:55 Severe earthquake continuing all nite.

All men called to stations. Legion emergency sounded, Detail of men to guard buildings and pull down chimneys. Many parapet walls fell and several persons injured by falling bricks.

10:00 PM approximately Sent Eng. #2 to Long Beach Capt. Honeycutt; Wells + Smith.

All men worked entire nite.

An earthquake, registering 6.3 on the Richter scale, struck the city of Long Beach and Southern California at 5:55 p.m. on March 10, 1933. One hundred and fifteen lives were lost and there was extensive property damage. Chief Crocker came on duty at 7:15 a.m. that morning. At 5:55 p.m., his shaky handwriting documents the earthquake: "Severe earthquake continuing all night. All men called to stations. Legion emergency sounded. Detailed men to guard buildings and pull down chimneys. Many parapet walls fell and several persons injured by falling bricks." Captain Honeycutt and crew were sent to Long Beach aboard Engine No. 2 to give aid where they could. (Courtesy of NBFD archives.)

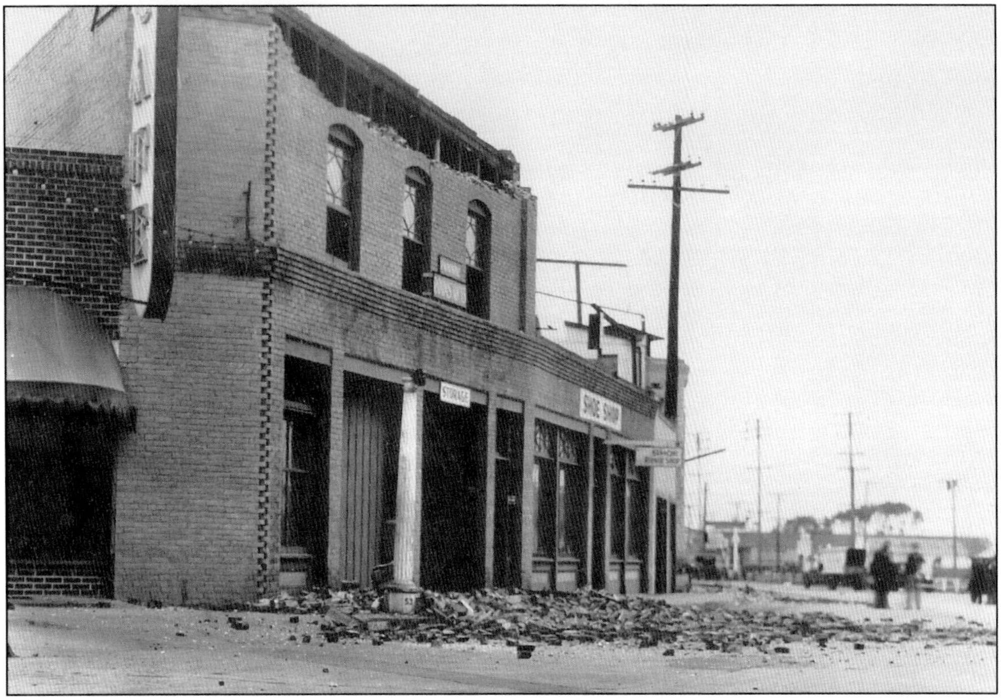

Five persons were hurt by falling bricks as they rushed out of this Newport Beach Café during the Long Beach earthquake. The majority of Newport residents, fearing a tidal wave, evacuated to high ground during the night. Every member of the department worked continuously for 24 hours, pulling down damaged chimneys and assisting citizens huddled on the surrounding bluffs. (Courtesy of NBFD archives.)

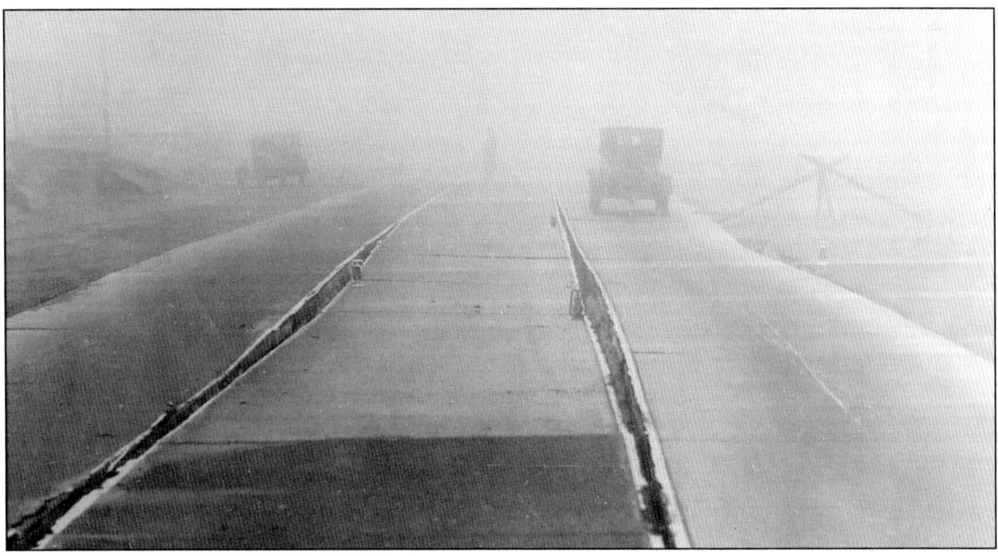

Pacific Coast Highway was damaged in several places. Surprisingly, gas leaks were few throughout the city and there were no fires. It was a long and adventurous night for Newport firefighters. Cryptically, Chief Crocker noted, "They [firefighters] also pulled many corks and turned down a few to warm the inner man." An exact interpretation of the chief's statement is best left to the reader's imagination. (Courtesy of NBFD archives.)

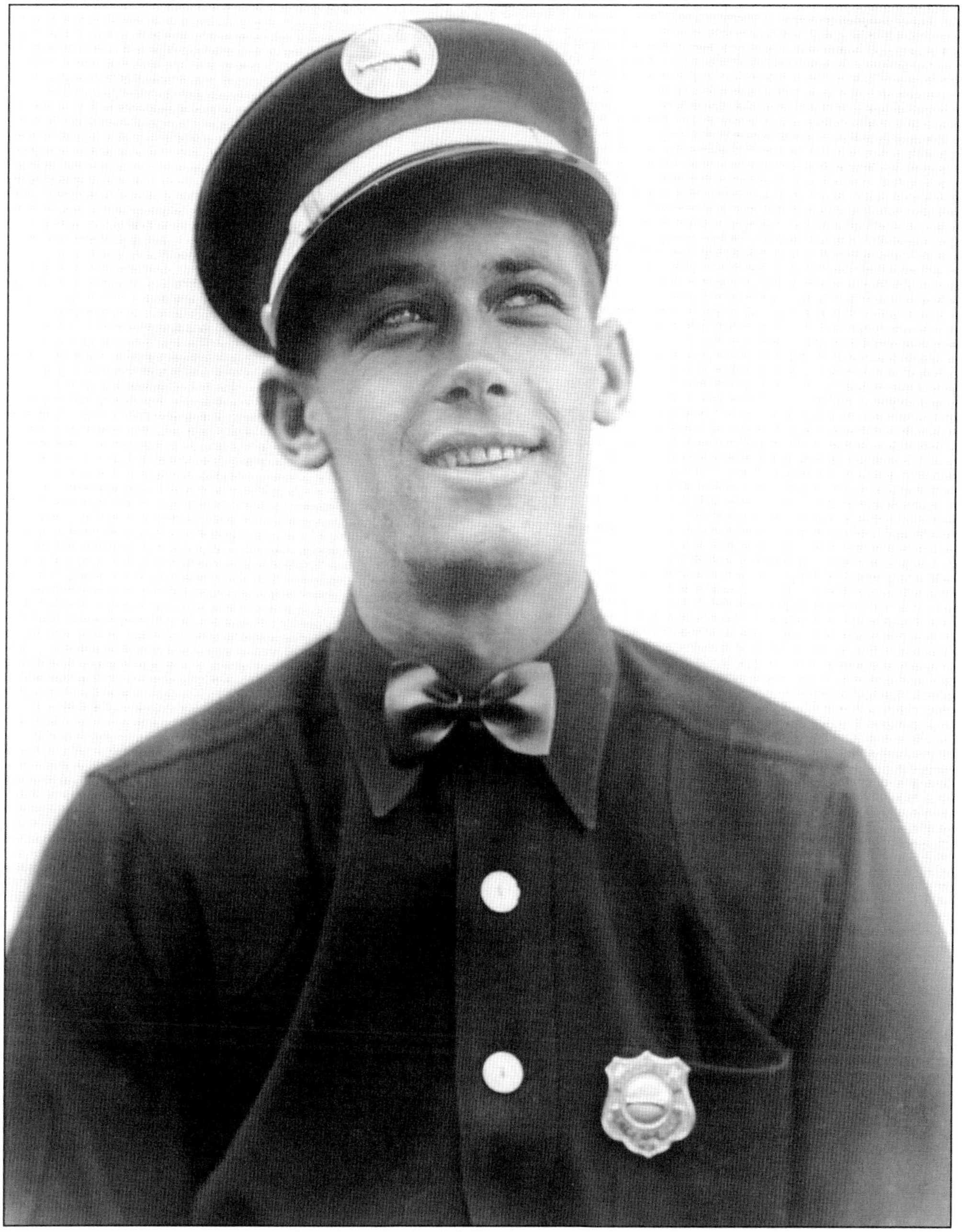

In 1933, construction was underway to rebuild and extend the west jetty at the harbor entrance. Railroad tracks were installed atop the jetty on giant timbers so that boulders could be delivered to the site. One night in September, huge waves tore some of these massive timbers off the jetty and pushed them toward nearby Balboa Pier. Onlookers feared that contact with the pilings could destroy the pier. Firefighters were called to the scene and launched a small boat in the rough seas. One by one, they quickly attached ropes to the errant timbers and pulled them to the beach. Orlando Briscoe, brother of Jan Briscoe, was attaching a rope to one 50-foot-long log when his clothing caught on a spike. Briscoe had no other option but to valiantly ride the timber to the beach through the huge surf, where he safely rejoined his comrades. (Courtesy Don Crocker.)

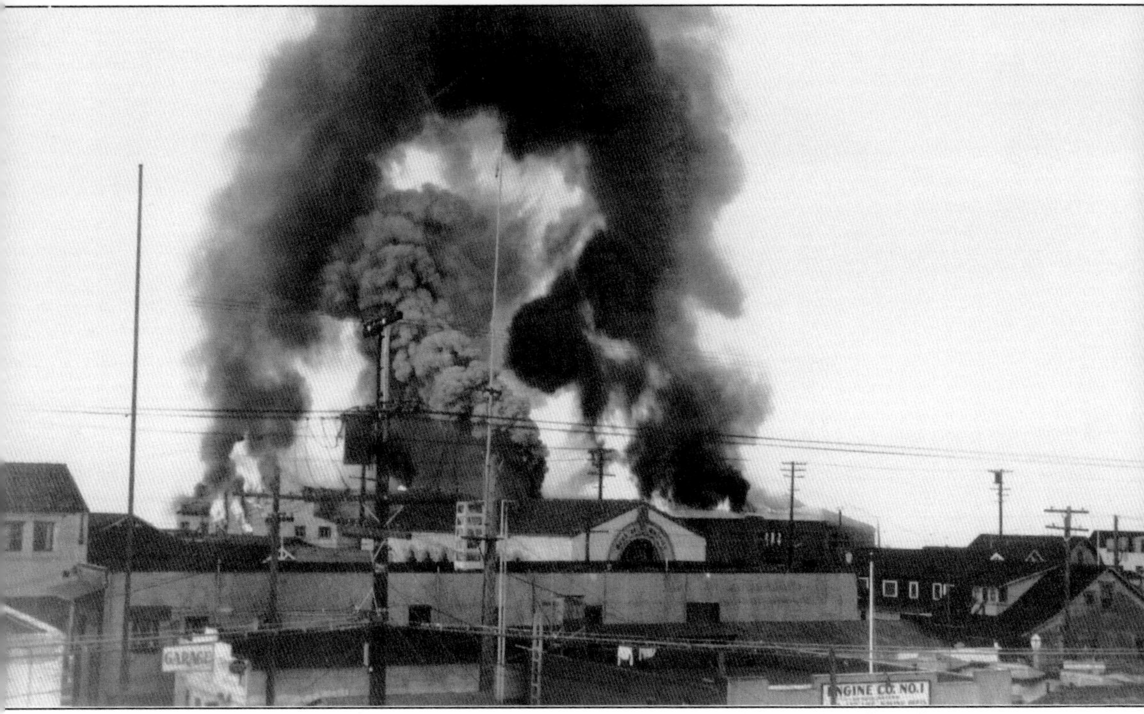

Early on the morning of January 27, 1935, the fire whistle pierced the calm, clear skies, summoning firefighters to one of the largest fires in Newport Beach history. Engines No. 1 and No. 2 were dispatched at 5:55 a.m. This was the view that greeted responding firefighters; the Rendezvous Ballroom was completely engulfed in flames. The Rendezvous, opened in 1928, was located on the beach in Balboa. It was a popular venue for listening and dancing to live music. Many NBFD social functions had been held at the Rendezvous over the years. Fire Station One, seen in the lower right foreground of this photograph, was only a short walk away. (Courtesy of NBFD archives.)

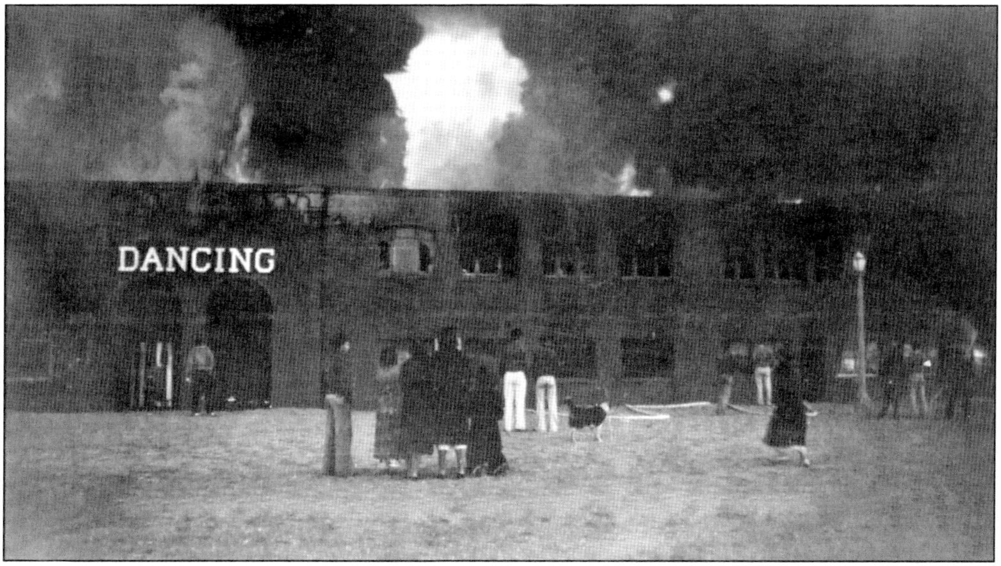

A discarded cigarette ignited a couch several hours after the ballroom had closed for the evening. When firefighters reached the scene, the Rendezvous was already a lost cause. Assistant chief Jan Briscoe's shift started at 7:00 a.m. that morning. He was on his way to work and could see the smoke and flames. He went directly to the fire, where he snapped these photographs before donning his turnout jacket and grabbing a hose line. (Courtesy of NBFD archives.)

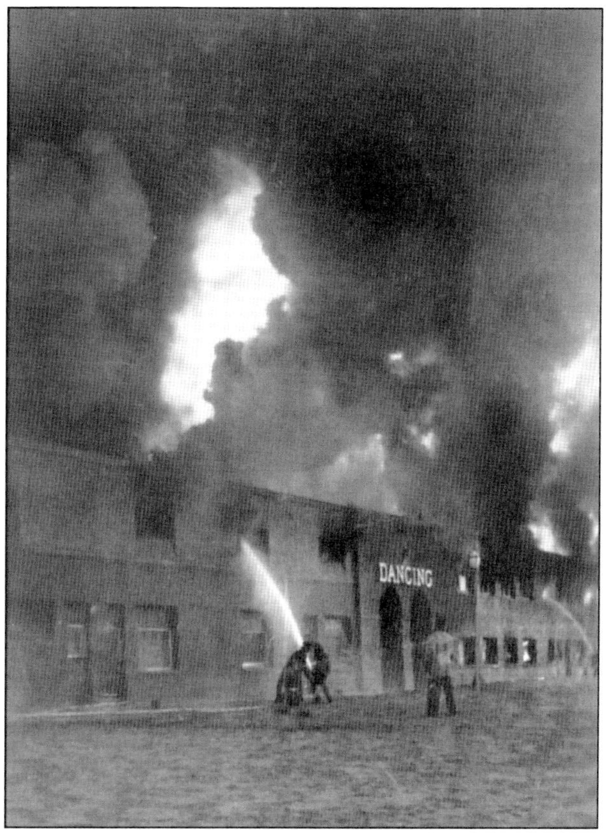

A voluminous amount of smoke lifted high into the sunny skies and could be seen from every corner of the city. In this Briscoe photograph, hose streams are being brought to bear on the walls of the ballroom. (Courtesy of NBFD archives.)

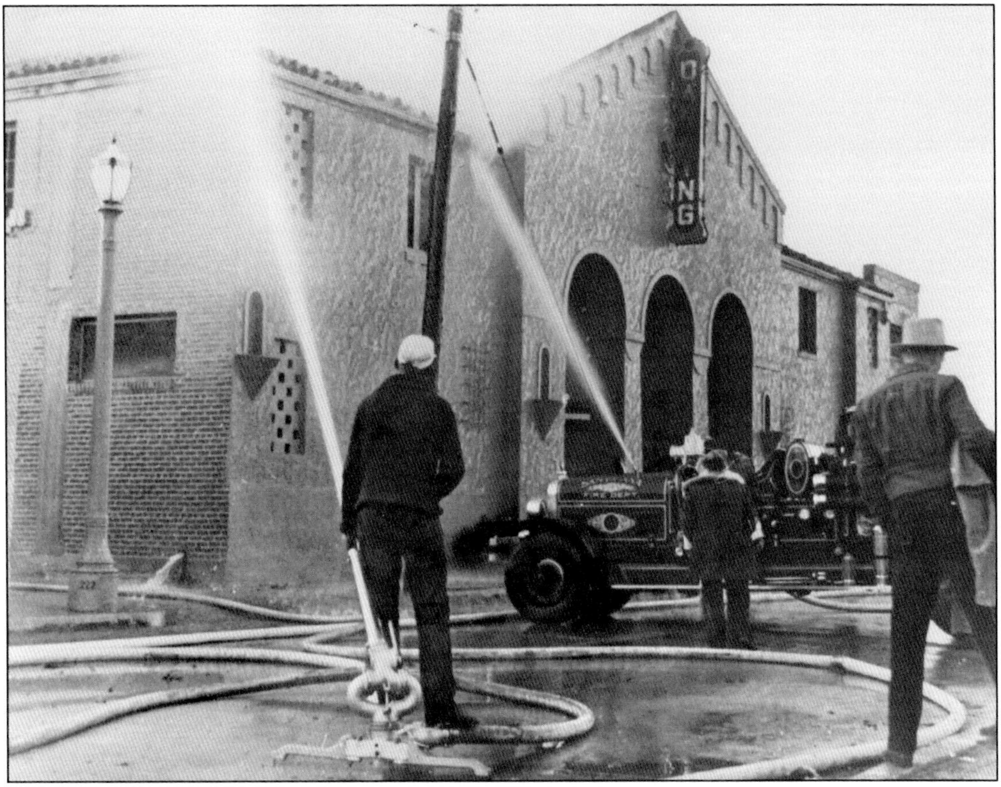

This view shows a Santa Ana Fire Department engine assisting in the extinguishment of the Rendezvous fire. Station One received calls from both the Huntington Beach and Santa Ana Fire Departments asking if assistance was needed. "Yes, hurry!" was the response. Engines from nearby Huntington and Santa Ana arrived shortly thereafter to help, a practice that would become commonplace at large fires in the future. (Courtesy of NBFD archives.)

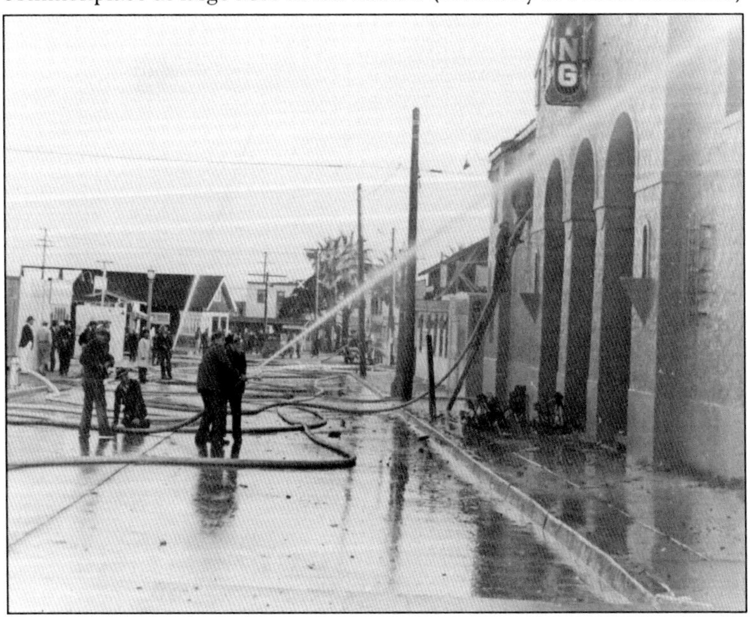

Because of the intense heat, firefighters were forced to fight the fire from outside the ballroom. Newspaper accounts stated that firefighters directed water upon the interior for five hours. (Courtesy of NBFD archives.)

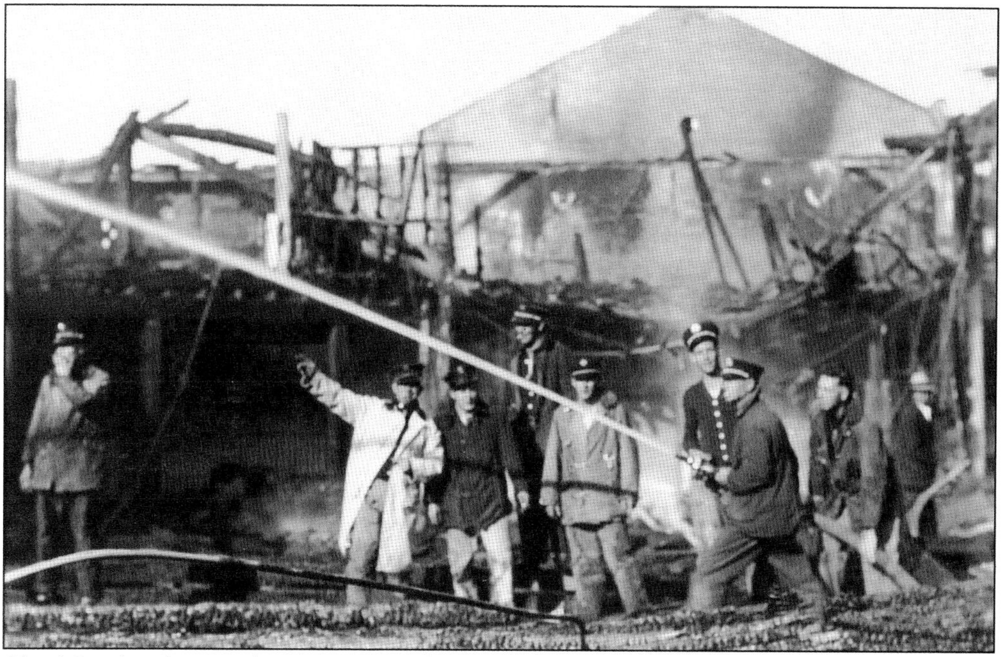

Chief Crocker directs a hose stream inside the ballroom ruins. The interior of the building had to be cooled and rendered safe before any kind of investigation could be undertaken. (Courtesy of NBFD archives.)

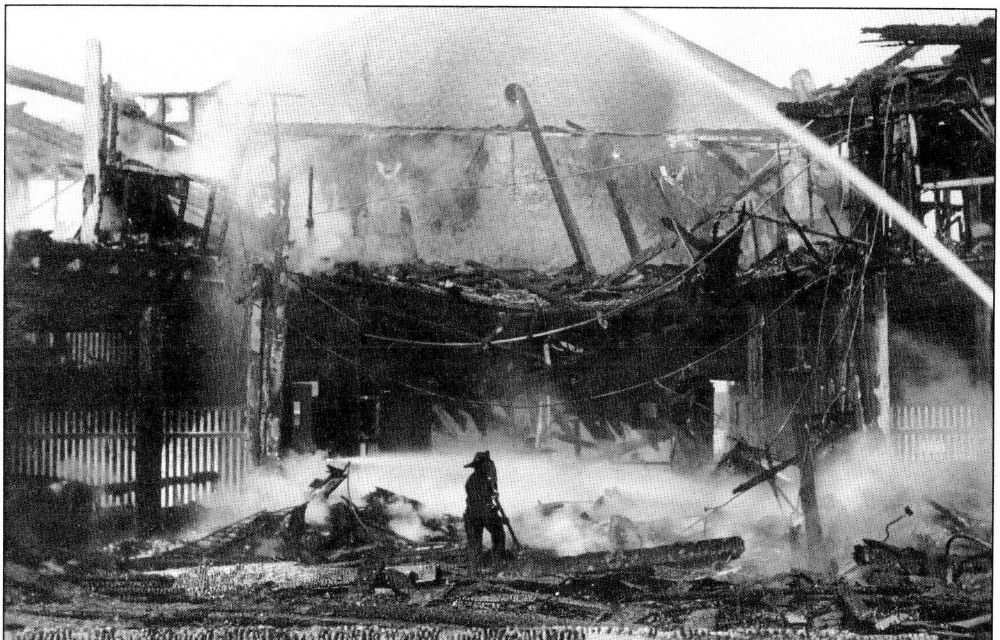

A silhouetted firefighter takes a breather as streams of water continue to extinguish hot spots inside the ballroom. Firefighters were fortunate that the roof collapsed as quickly as it did. Sending firefighters into the Rendezvous while the roof was burning overhead would have been extremely dangerous. Firefighters remained mindful of the possibility of the brick walls collapsing inward. (Courtesy of NBFD archives.)

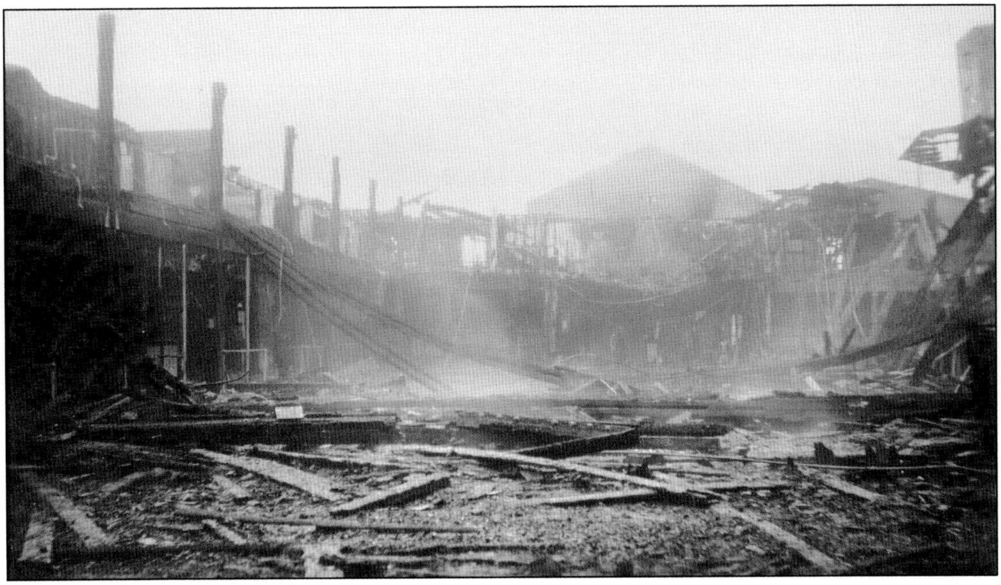

The roof of the ballroom collapsed 20 minutes after the fire was reported, leaving only the exterior walls standing. These walls would be all that remained of the Rendezvous Ballroom at the end of the day. Its owners reported a loss of $100,000. Chief Crocker's estimate was considerably lower. (Courtesy of NBFD archives.)

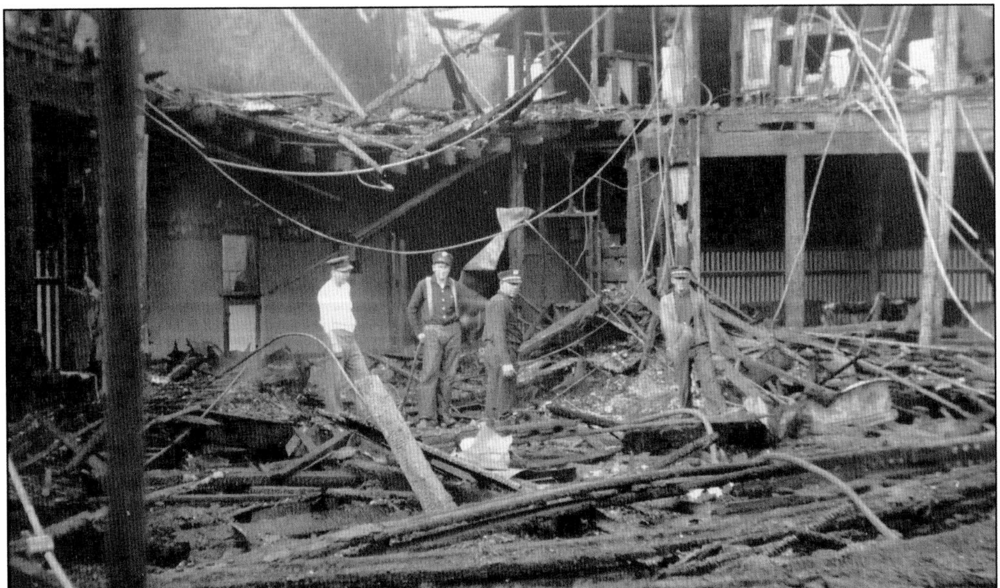

The Rendezvous Ballroom fire was the largest in Newport Beach's history to date. Indeed, Chief Briscoe would later call it the largest of his 44-year career. Remarkably, rebuilding the ballroom started immediately after the fire. The Rendezvous would be open for business, complete with a new spring dance floor, two months later. (Courtesy of NBFD archives.)

One can only wonder what was going on behind Station Two the day this photograph was taken. Nowadays firefighters practice resuscitation and CPR techniques using hi-tech, lifelike manikins. Perhaps this fireman's companion was an early example of such a training aid. (Courtesy of NBFD archives.)

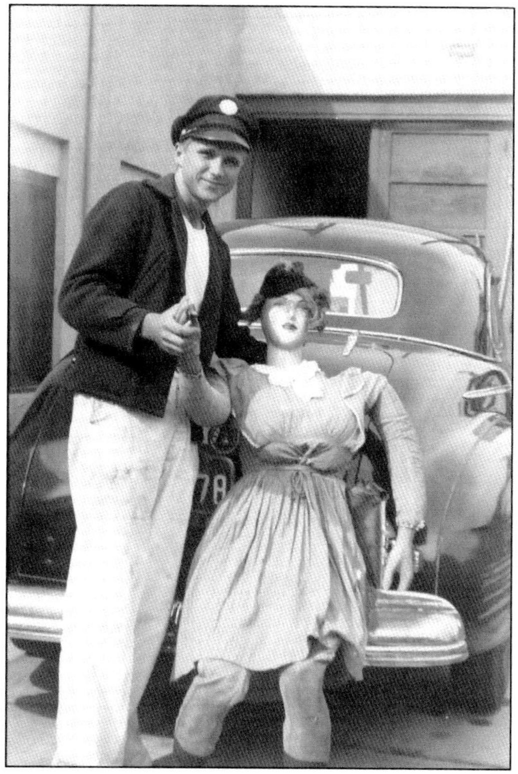

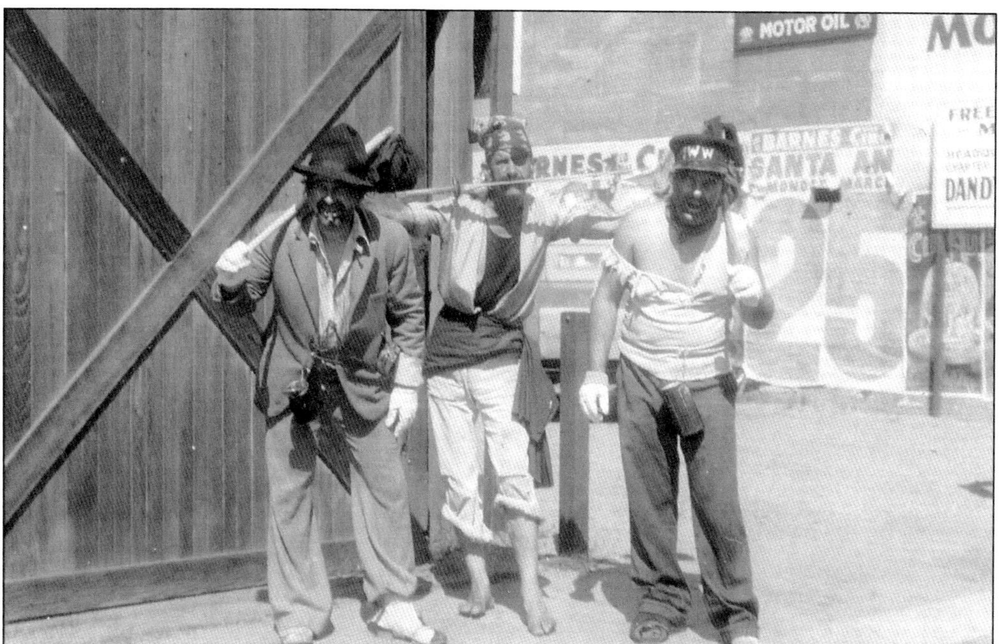

At times, a firefighter's shift can be subject to periods of high stress. These three firefighters found a moment of off-duty relief by clowning around behind the station; the chief is on the left. Behind them, posters announcing the arrival of the Al G. Barnes Circus in nearby Santa Ana cover the walls. (Courtesy of NBFD archives.)

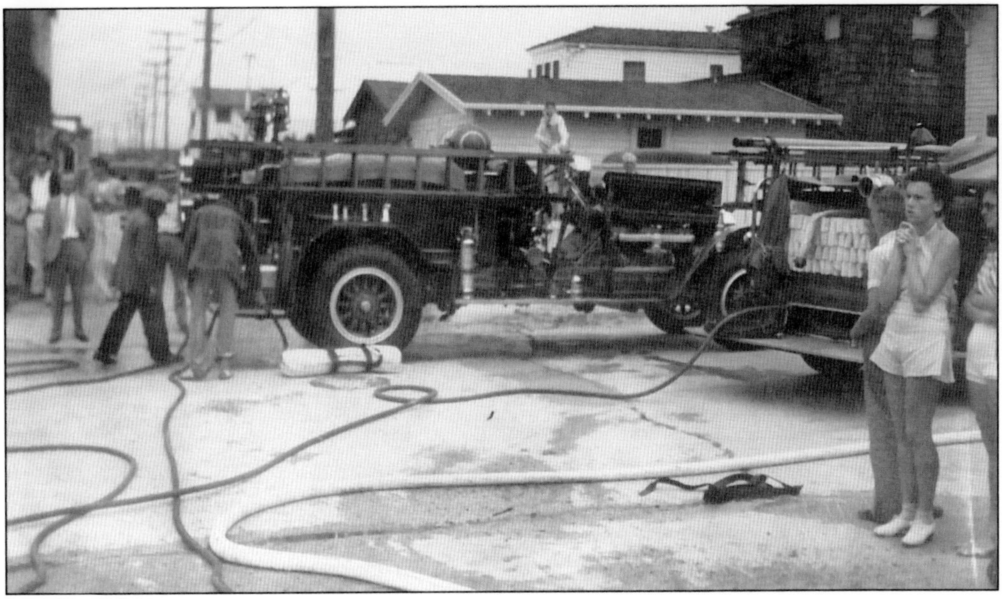

This photograph is one of the first to show NBFD fire engines in action at a fire. In May 1935, the summer home of the Raidman family, who lived in Los Angeles, caught fire. A faulty gas oven was the culprit. In this view, Engines No. 1 and No. 2 provide small-diameter rubber hoses called leader lines for extinguishing the fire. (Courtesy of NBFD archives.)

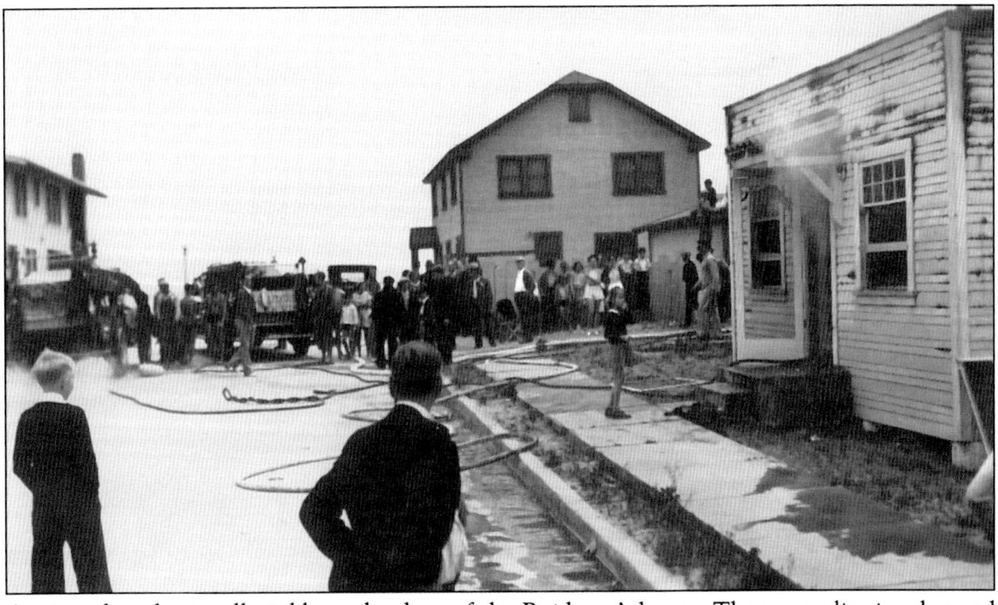

A wisp of smoke is still visible at the door of the Raidman's house. The ocean lies just beyond the two engines. The loss was put at $700. Fire loss for the year 1935 was officially recorded at over $40,000—the majority of that loss attributable to the Rendezvous Ballroom fire in January. (Courtesy of NBFD archives.)

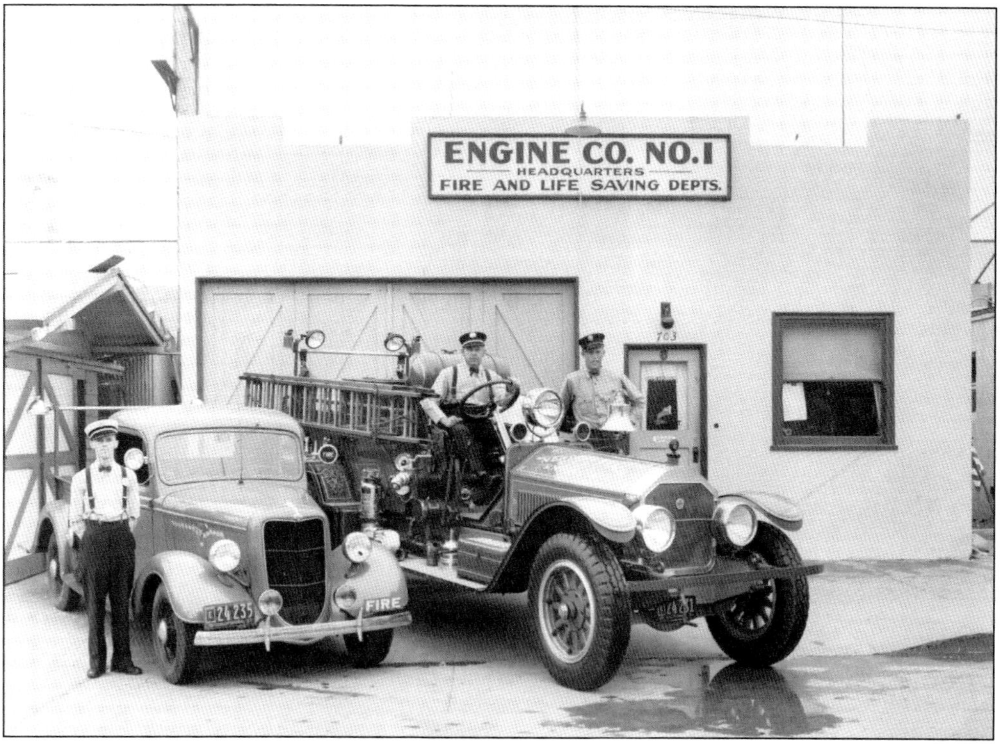

Small changes had been made by 1937. Station One now had a new sign gracing its facade. The lifesaving corps has now been acknowledged. A new Ford pickup truck has replaced the Ford Model A. Jan Briscoe and Perce Pellett are wearing that most traditional of firefighter's gear—red suspenders. (Courtesy of NBFD archives.)

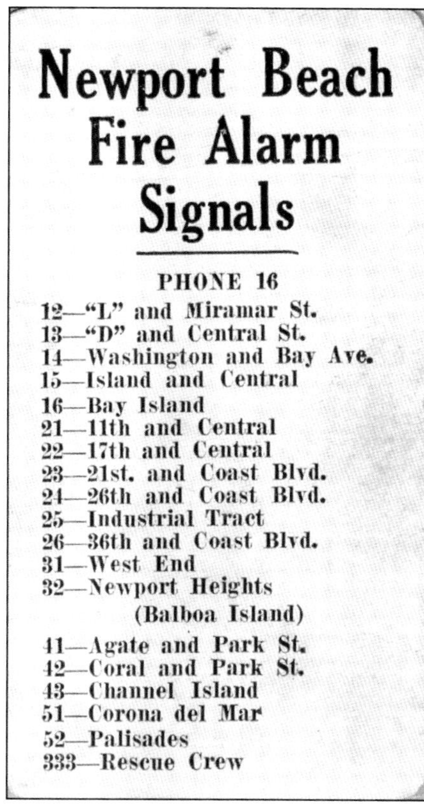

Newport Beach Fire Alarm Signals

PHONE 16

12—"L" and Miramar St.
13—"D" and Central St.
14—Washington and Bay Ave.
15—Island and Central
16—Bay Island
21—11th and Central
22—17th and Central
23—21st. and Coast Blvd.
24—26th and Coast Blvd.
25—Industrial Tract
26—36th and Coast Blvd.
31—West End
32—Newport Heights

(Balboa Island)

41—Agate and Park St.
42—Coral and Park St.
43—Channel Island
51—Corona del Mar
52—Palisades
333—Rescue Crew

This card was issued to all firefighters. When the whistles blasted out an alarm, firefighters could identify the approximate location and respond directly to the fire. (Courtesy of NBFD archives.)

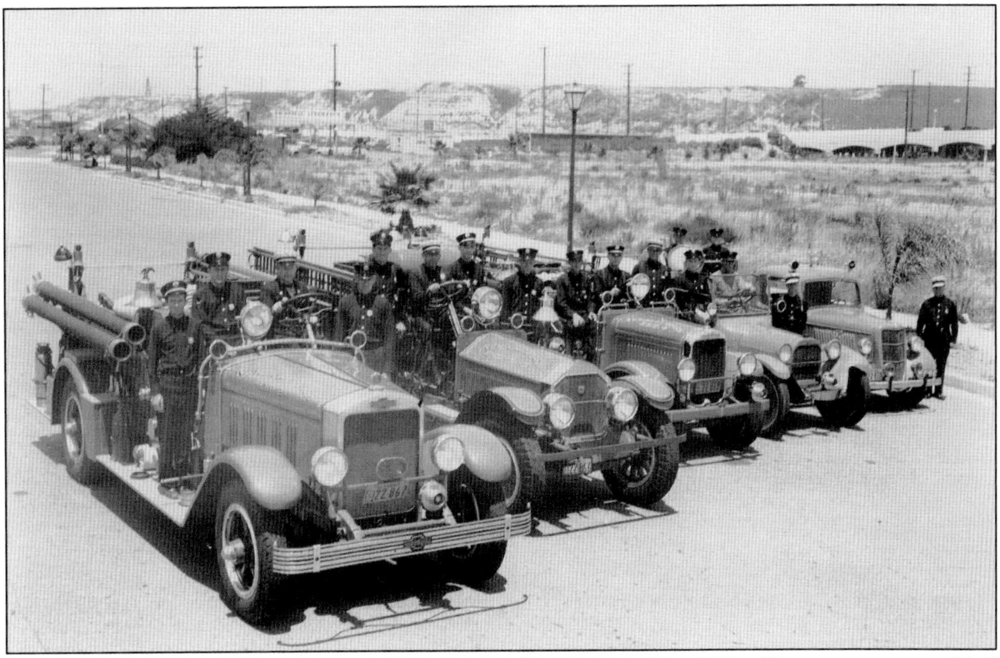

In 1938, ten years after the first group photograph, the Newport Beach Fire Department assembled once again in an area now known as Lido Village. The Squad, second from the right, is still going strong. (Courtesy of Don Crocker.)

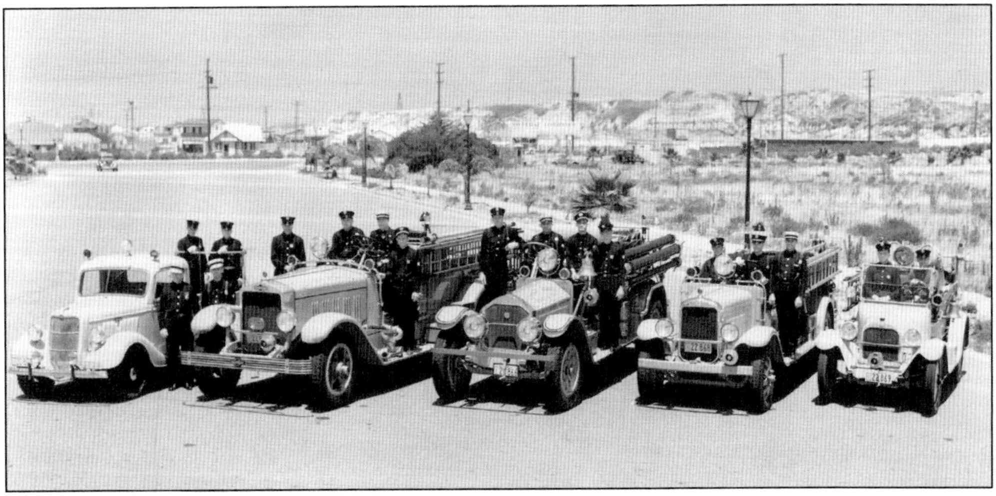

When these photographs were taken, the onset of World War II was only a few years away. The area in this view would soon be surrounded by boatyards building minesweepers and doing conversion work for the U.S. Navy. Chief Crocker would do his part by monitoring meteorological forecasts for the civil defense effort. (Courtesy of Don Crocker.)

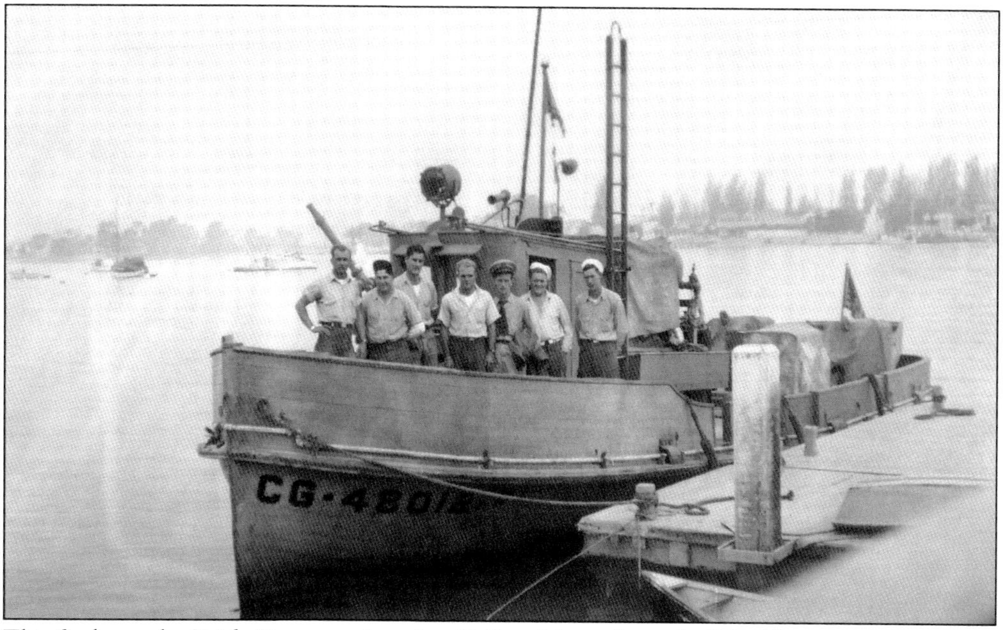

This fireboat, the *Bright II*, was stationed at Collins Isle at the end of World War II. The *Bright II* was operated by the U.S. Coast Guard, which still maintains a base and a brand-new cutter, the *Narwhal*, inside Newport Harbor. (Courtesy of NBFD archives.)

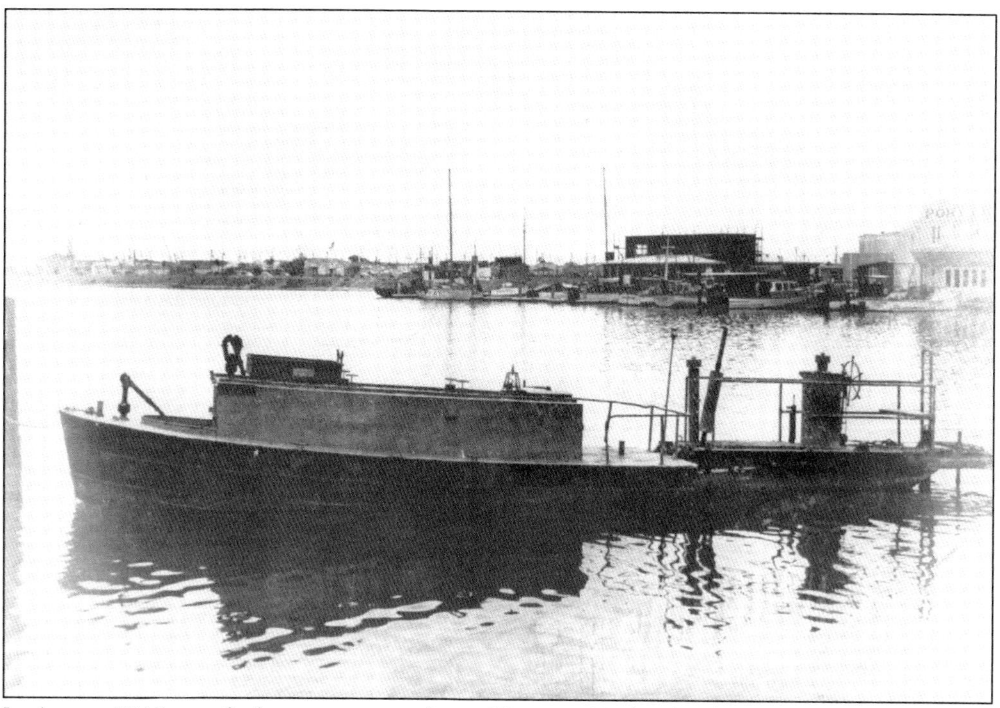

In August 1946, two fireboats were towed into Newport Harbor by the U.S. Coast Guard. Each had two monitor nozzles and was maneuvered by a sea-going mule. It is unknown how long they stayed in the harbor or if they were ever used. (Courtesy of NBFD archives.)

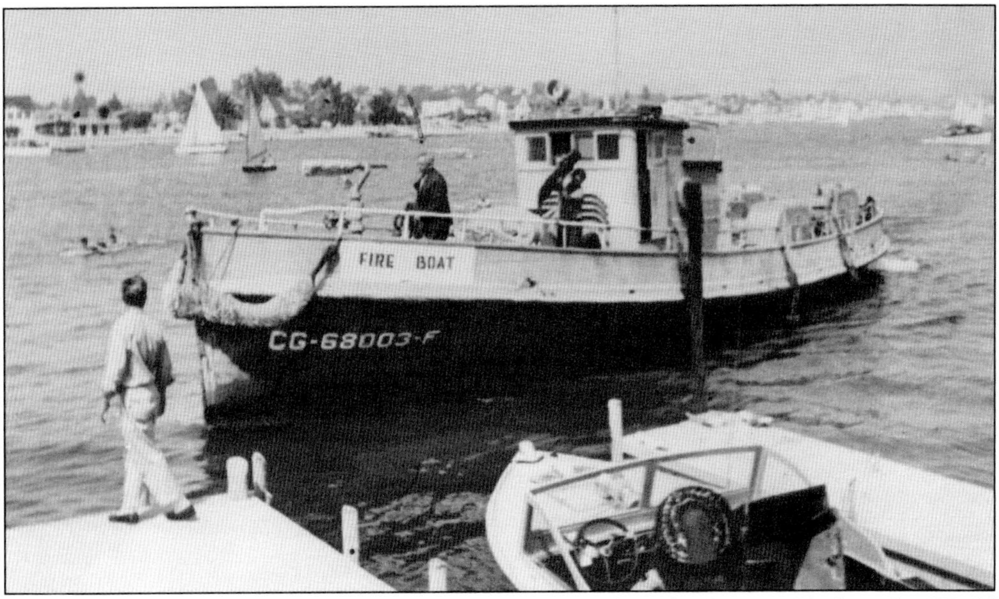

Chief Crocker takes a ride on one of the fireboats stationed inside Newport Harbor after World War II. These boats were operated by the U.S. Coast Guard. The use of fireboats was first mentioned by city trustee H. L. Sherman in 1920, but his suggestion fell on deaf ears. (Courtesy of NBFD archives.)

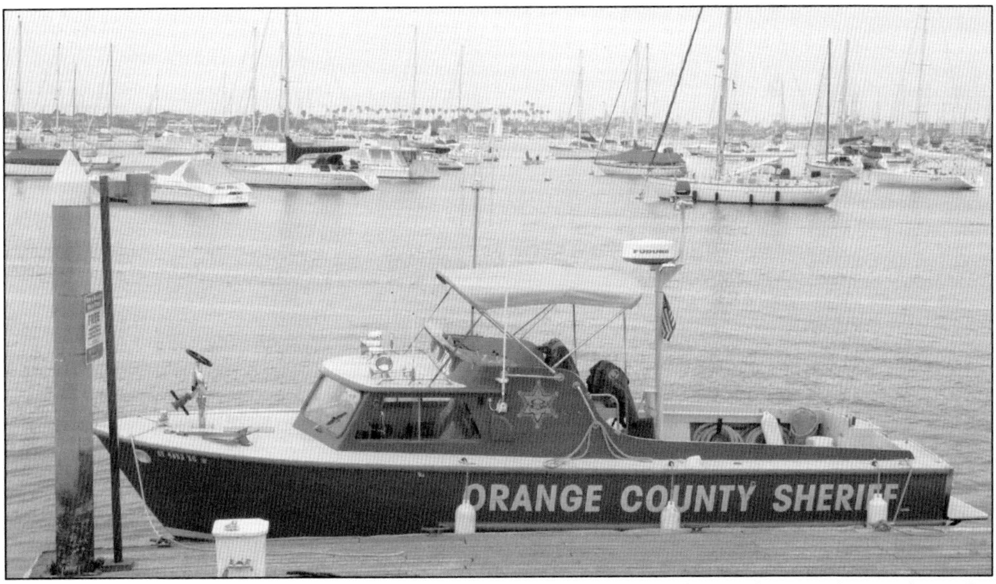

Today the Orange County Sheriff's Department operates smaller fire and rescue boats along the coast and throughout the harbor. Fireboats respond with the NBFD to all boat and waterfront structure fires. They are essential for accessing the side of a waterfront home not approachable from the street and vessels moored along lengthy docks. (Author's collection.)

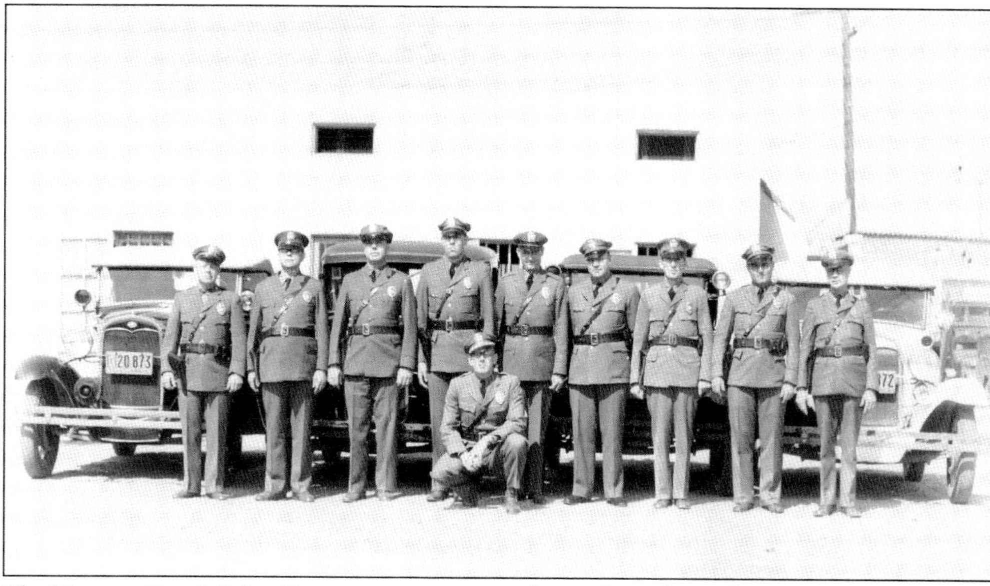

The Newport Beach Police Department experienced growth similar to that of the fire department. By the early 1930s, the force consisted of 10 officers and four squad cars. This was a far cry from the lone patrolman at the turn of the century who kept one eye on a light mounted on a pole by the Newport Pier. If the telephone operator received an emergency call, she would pull a switch that would illuminate the red light, alerting the officer—provided he was looking. (Courtesy of Don Crocker.)

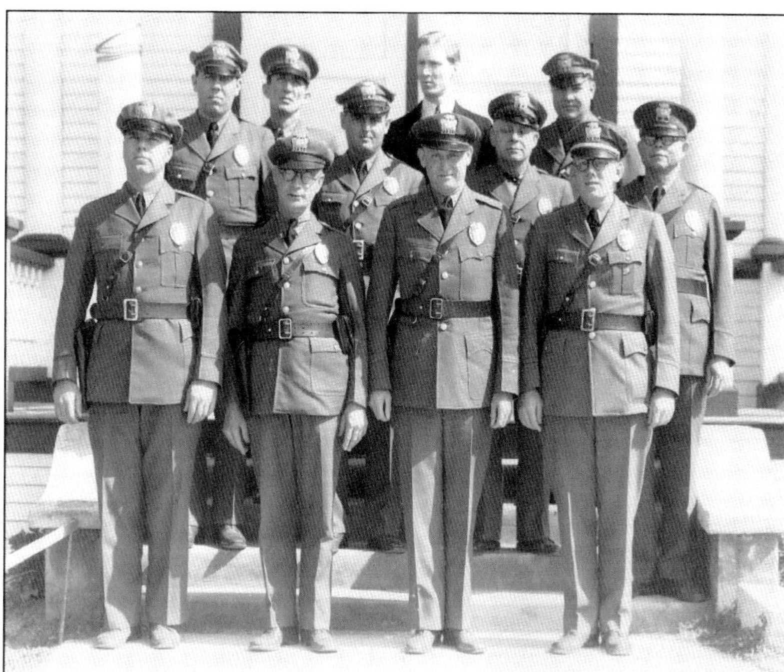

Members of the Newport Beach Police Department stand on the steps of the original city hall, a former schoolhouse near the Newport Pier. (Courtesy of NBFD archives.)

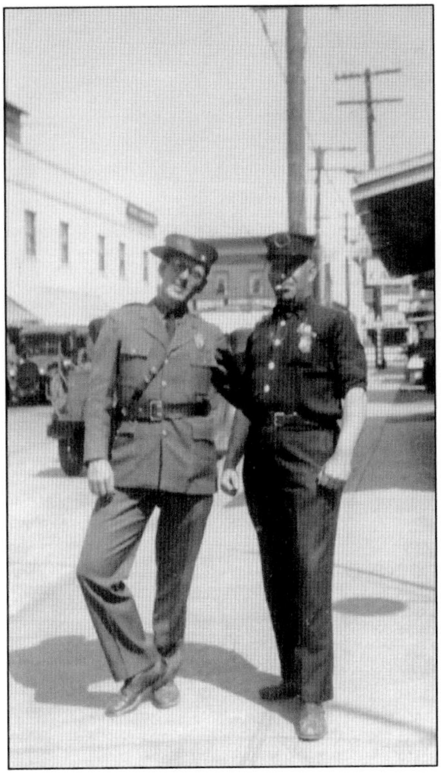

Chief Crocker and police chief Rowland Hodgkinson strike a jaunty pose in front of Fire Station One. The Newport Beach Fire and Police Departments have always enjoyed a close working relationship. Chief Crocker and Chief Hodgkinson were appointed to their respective positions the same year, and they retired together in 1952. (Courtesy of NBFD archives.)

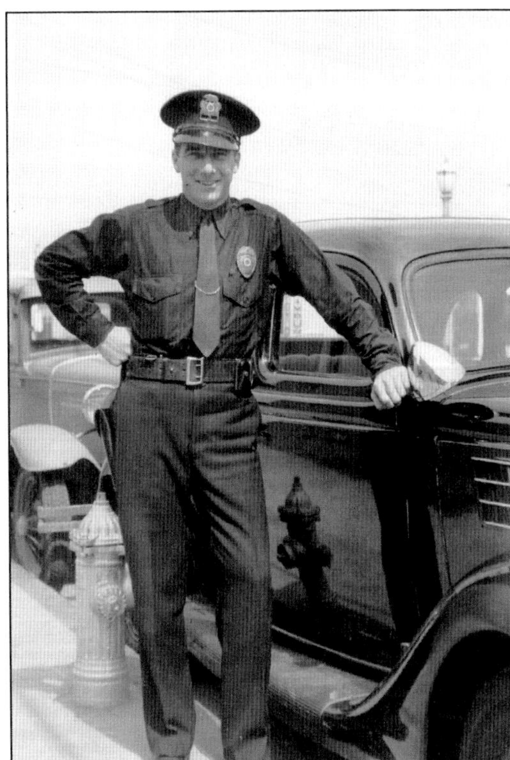

Cuba Morris was a paid-call firefighter who, when offered an opportunity to join the Newport Beach Police Department, jumped at it. Morris is shown here, needling his former fire mates by blocking a fire hydrant with his squad car. (Courtesy of Don Crocker.)

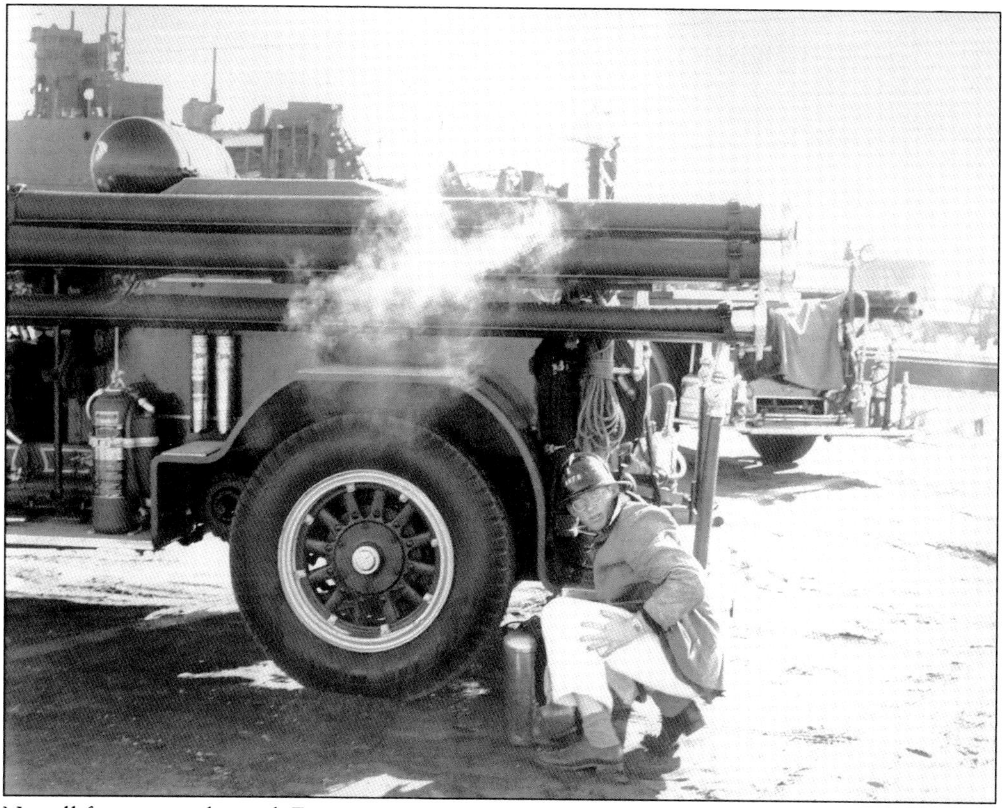

Not all fires go as planned. Engine No. 2 responded to a boat fire in May 1948, and firefighters quickly extinguished the flames. Engine No. 1 arrived to help, only to notice that Engine No. 2's brakes were on fire. In this view, a firefighter has just used a fire extinguisher to quell the brake fire. (Courtesy of NBFD archives.)

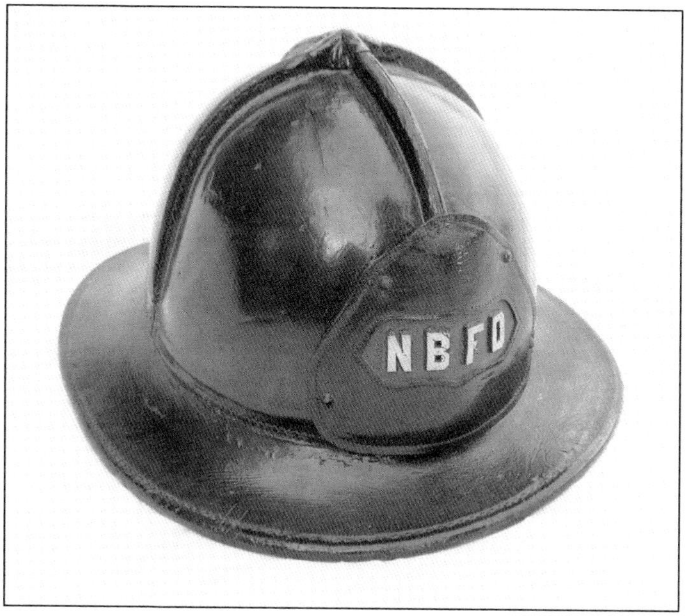

This style of fire helmet, made of hard leather, was worn by Newport Beach firefighters during the 1940s. This particular helmet was given by Chief Leo Love to battalion chief Milt Meehan shortly before Chief Love passed away. Chief Meehan bequeathed it to the NBFD in honor of his friend. (Courtesy of NBFD archives; helmet gift of Milt Meehan.)

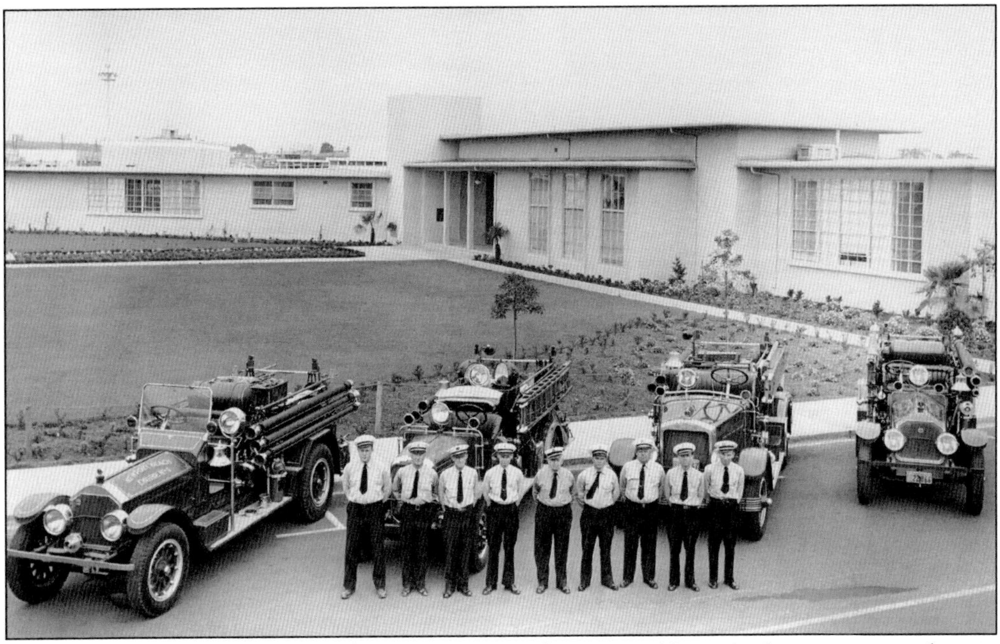

A group photograph of the fire department was taken every 10 years; this view was taken in 1948. Newport Beach City Hall lies behind them. The American LaFrance engine on the left was purchased around 1938 from the Los Angeles City Fire Department. It was sold in the early 1960s and has since been restored and used in parades by its new owner in Oregon. (Courtesy of NBFD archives.)

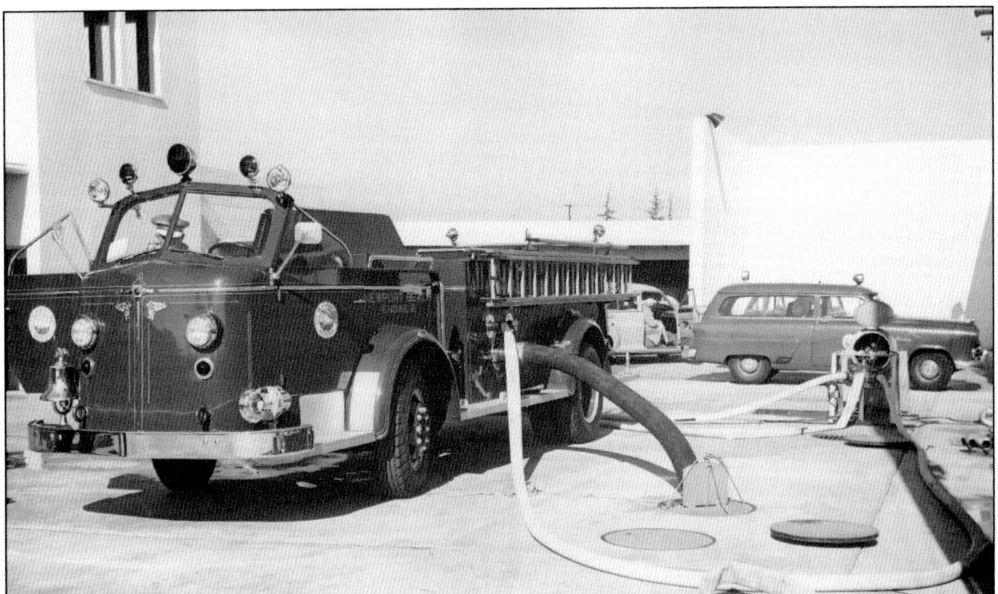

Newport Beach took delivery of a new American LaFrance 700 Series pumper in 1952. This fire engine was a radical departure from previous models. Its cab-over feature put the crew forward of the engine, an innovation that would influence all subsequent fire engine design. (Courtesy of NBFD archives.)

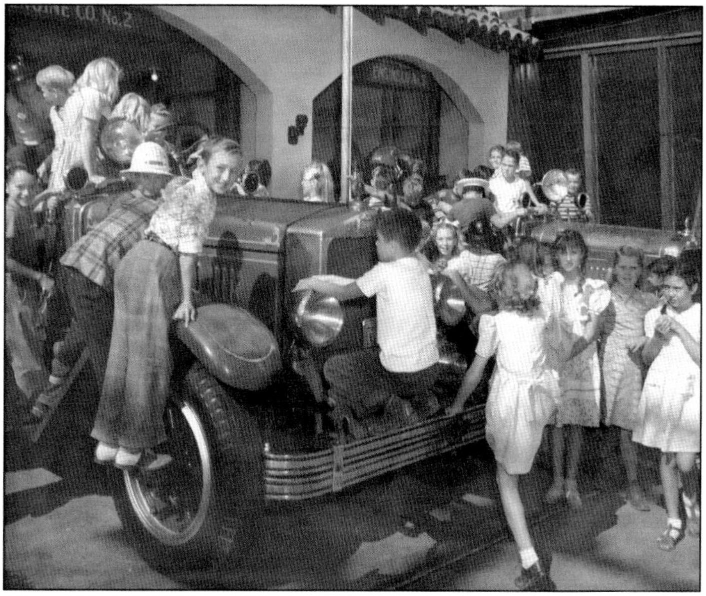

Field trips to the local fire station have always been a favorite for school kids. In this photograph, children visiting Station Two are making short work of a rookie firefighter's morning polish job. It is ironic that, at the end of its useful fire-service life, this 1931 LaFrance would be partially buried in the sand in a neighborhood park, where it would be climbed and played on by children until it was dug out and scrapped in 1971. (Courtesy of NBFD archives.)

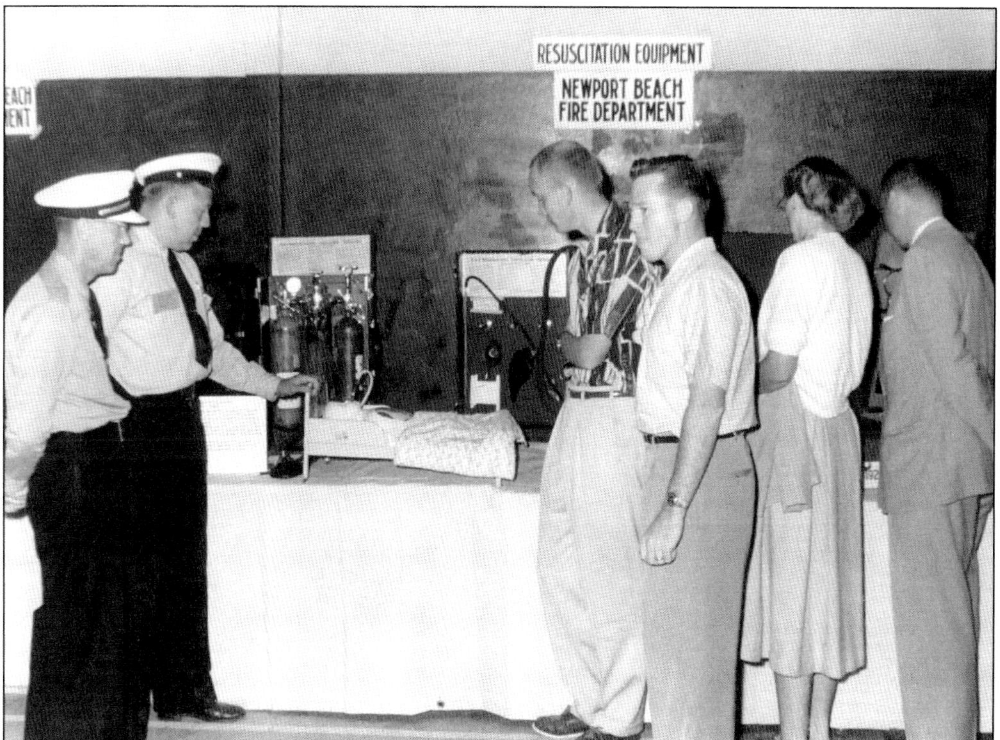

Fire prevention and public education have always been priorities of the Newport Beach Fire Department. In this photograph, Chief Briscoe and fire marshal Bill Noller display and demonstrate the department's latest resuscitation equipment to curious citizens. (Courtesy of NBFD archives.)

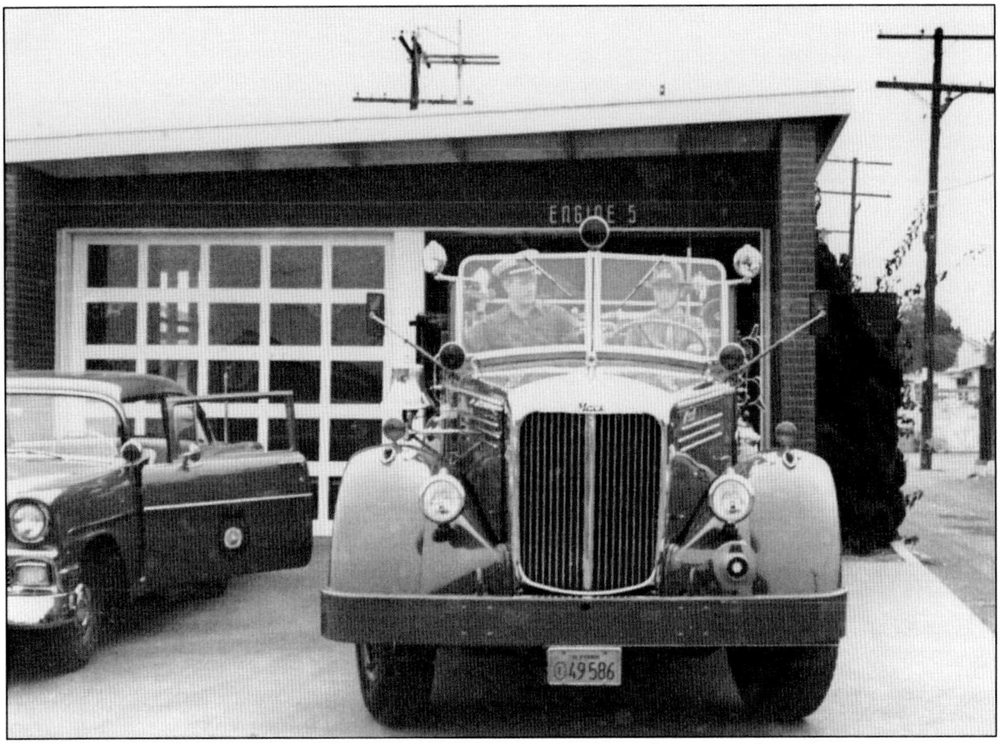

Corona del Mar finally got its own fire station in 1949. Prior to that year, this growing community got by with help from Balboa Island and one of the original hose reels purchased in 1910. Today Station Five covers a large area, including the posh Newport Coast expanse between Newport and neighboring Laguna Beach. (Courtesy of NBFD archives.)

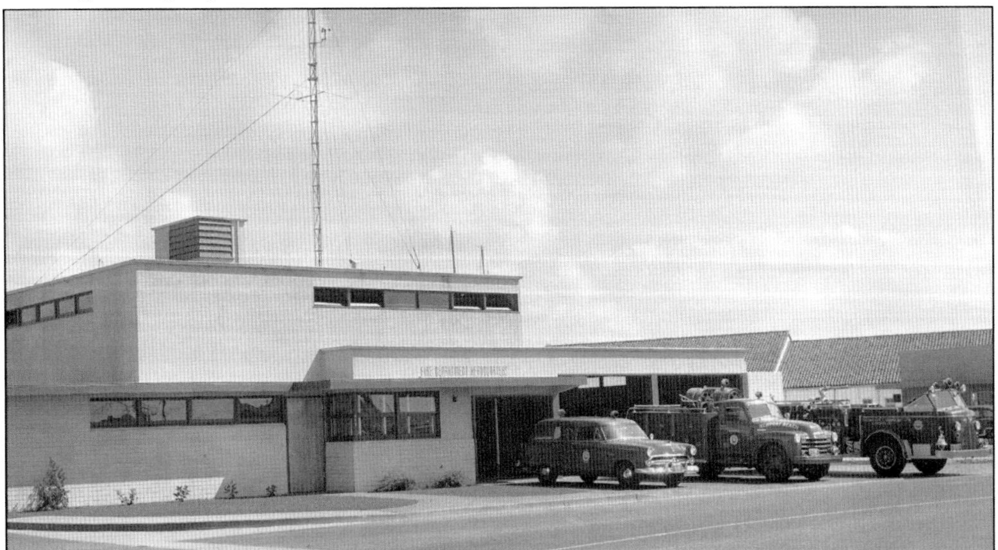

A new headquarters station was opened adjacent to city hall in 1953. Fire engines were getting larger and manpower was increasing. Old Station Two was replaced by this larger and more conveniently located fire station. The fire chief, fire prevention, and dispatch all maintained offices downstairs; firefighters lived upstairs. (Courtesy of NBFD archives.)

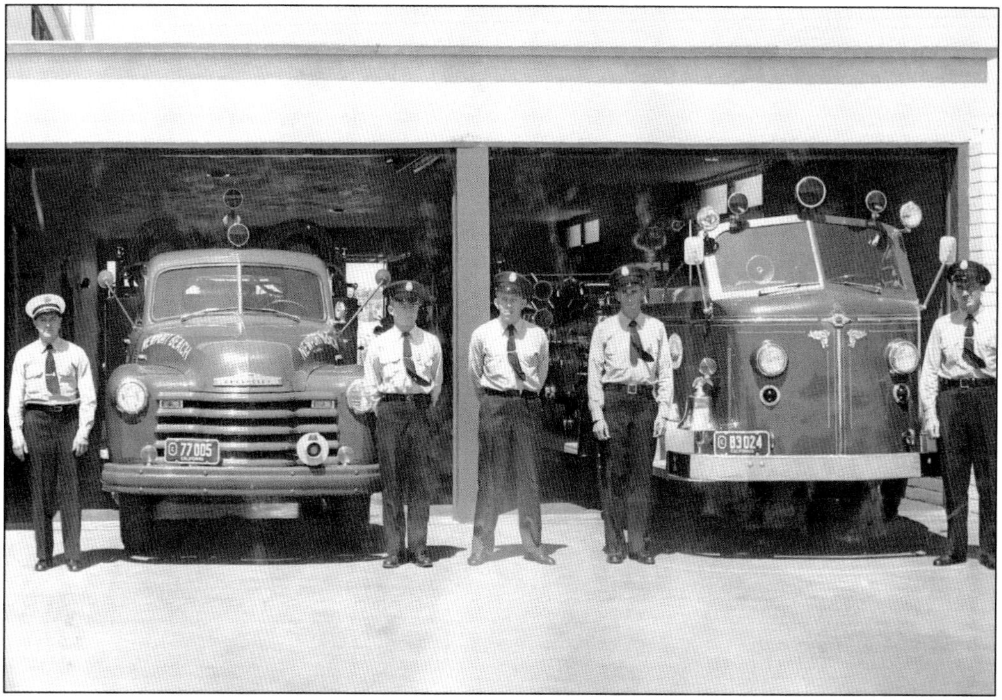

Six firefighters worked 24-hour shifts at Station Two. The 1952 American LaFrance and Chevrolet Squad filled the apparatus room. The addition of dedicated administrative offices downstairs signaled a new era in Newport Beach fire operations. (Courtesy of NBFD archives.)

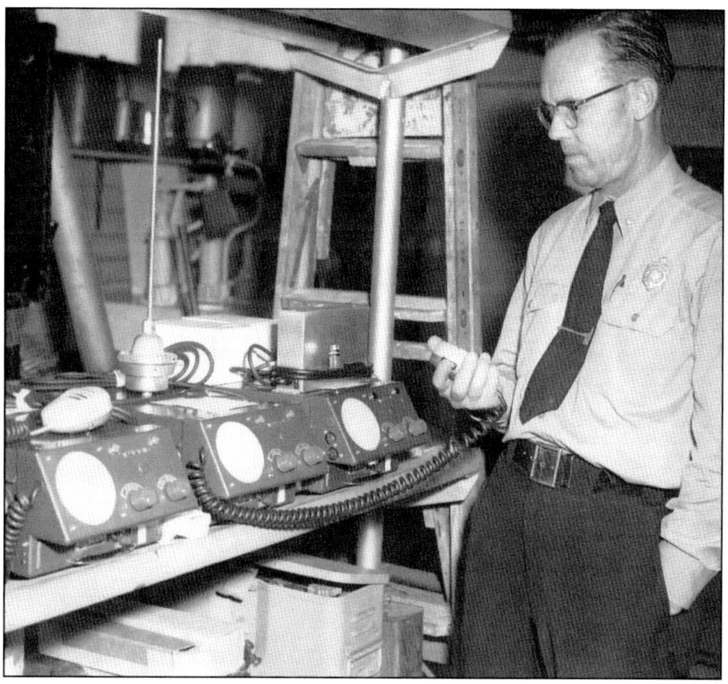

Chief Briscoe checks out three new radios purchased to facilitate communications between fire stations. Firefighters had previously relied on telephones and a telephone operator. Firefighters in the field could now communicate with the station. Radios were becoming increasingly important in the fire service. Today every firefighter carries a radio. (Courtesy of Richard Koehler.)

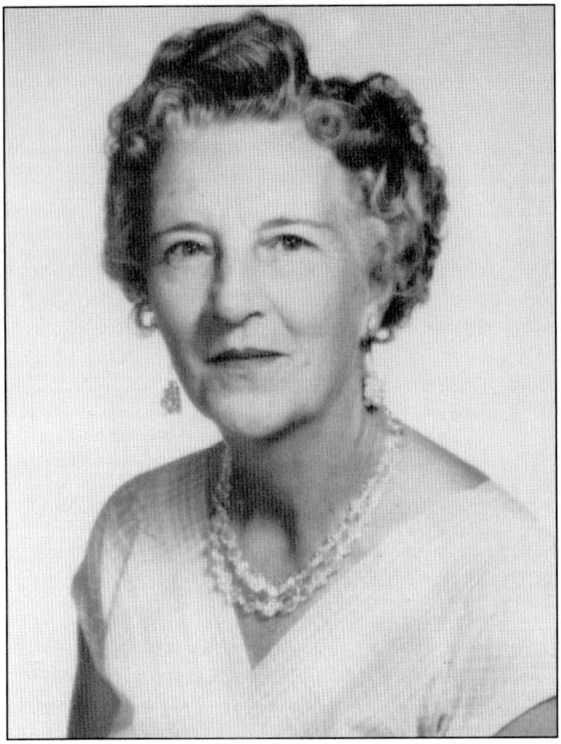

Fern Pellett, wife of assistant chief Perce Pellett, was Newport's telephone operator and, as such, the department's first fire dispatcher. In the late 1920s, emergency calls were relayed by the operator to Station One, where fire personnel would then initiate the proper response. (Courtesy of NBFD archives.)

Firefighter Dick Duncan acts out his mock frustration with the new fire department communication system for the camera. Efficient communication would become a cornerstone for effective fire department operations. (Courtesy of NBFD archives.)

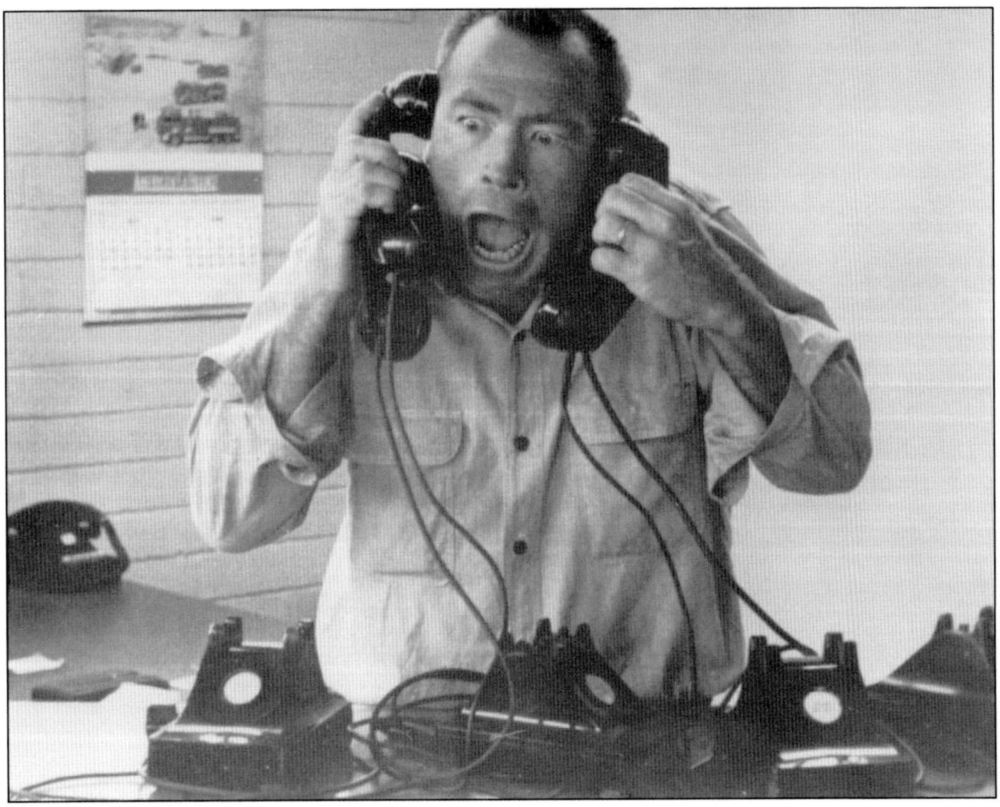

In 1953, a dispatch console was installed at Station Two. Two dispatchers, Jack Reimer and Hugh McMillan, were hired to work 12-hour shifts. In this view, Hugh McMillan, news reporter, local historian, author, and lifelong resident of Newport Beach, receives an emergency call and notifies the appropriate unit via radio. (Courtesy of Richard Koehler.)

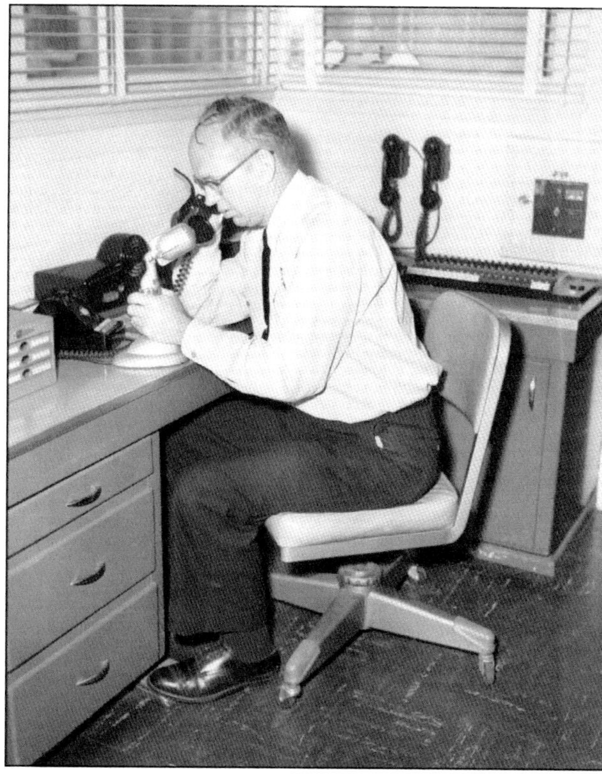

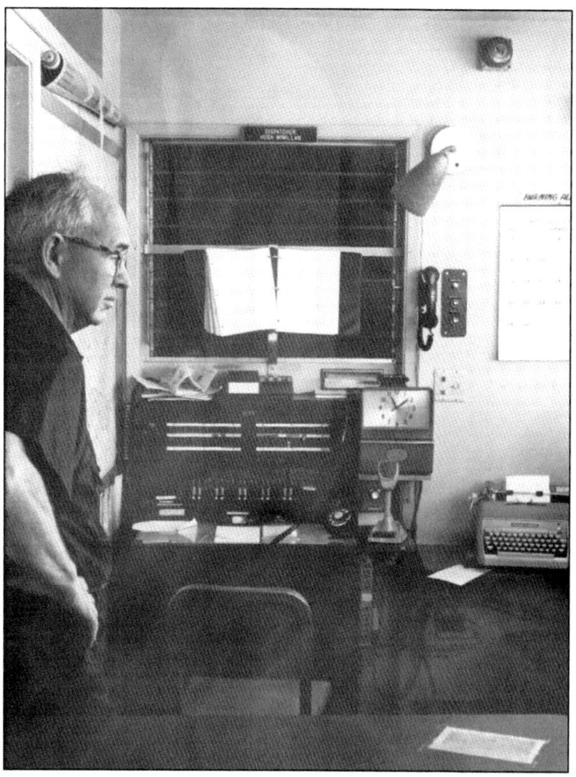

In 1973, Hugh McMillan was still at his post. A few years later, the Newport Beach Police Department took over the dispatching of fire units. Today Newport Fire receives alarm information via pagers, on-board computers, and radios from a state-of-the-art facility in Anaheim, California. (Courtesy of NBFD archives.)

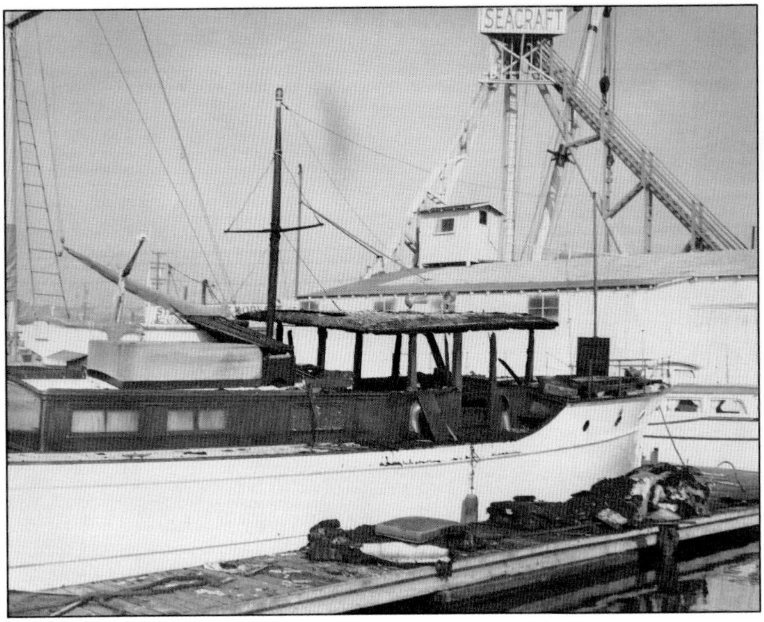

Newport Harbor is one of the largest pleasure boat harbors on the West Coast. Boat fires have presented firefighters with unique challenges, from the burning of eight boats and the loss of one man's life in 1928, to the destruction of the 100-foot *Mikey I* in 2006. The *Yankee D*, shown here, was destroyed by fire in 1948. (Courtesy of NBFD archives.)

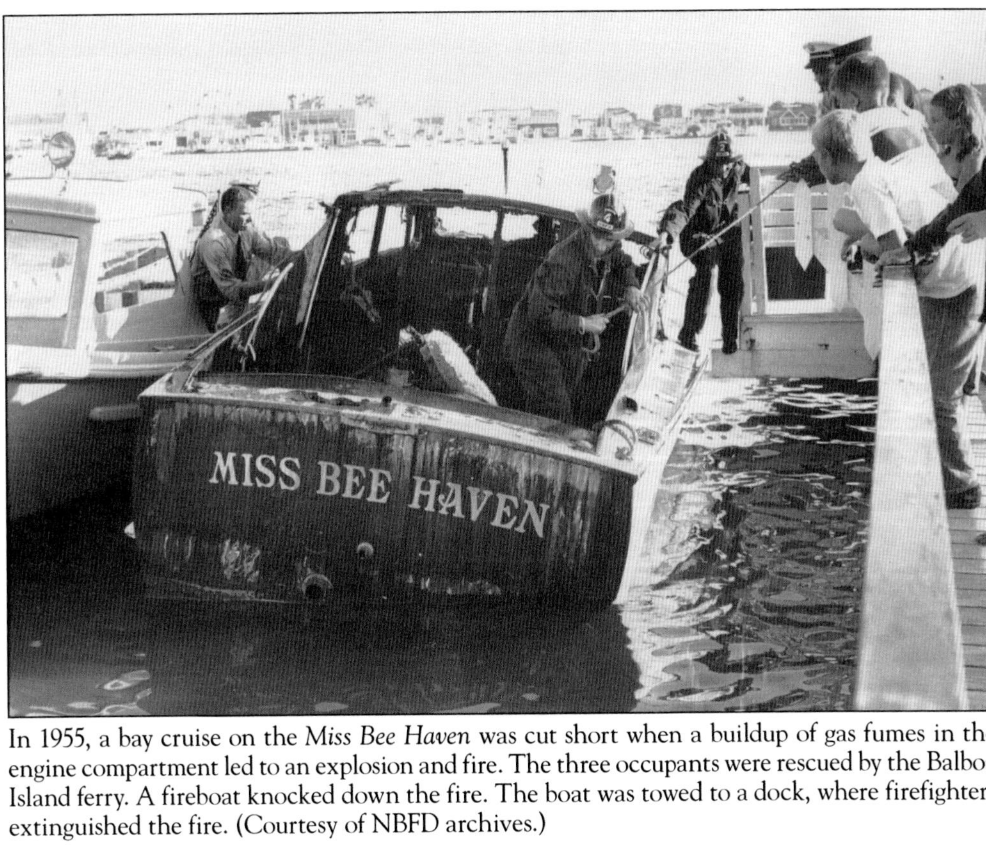

In 1955, a bay cruise on the *Miss Bee Haven* was cut short when a buildup of gas fumes in the engine compartment led to an explosion and fire. The three occupants were rescued by the Balboa Island ferry. A fireboat knocked down the fire. The boat was towed to a dock, where firefighters extinguished the fire. (Courtesy of NBFD archives.)

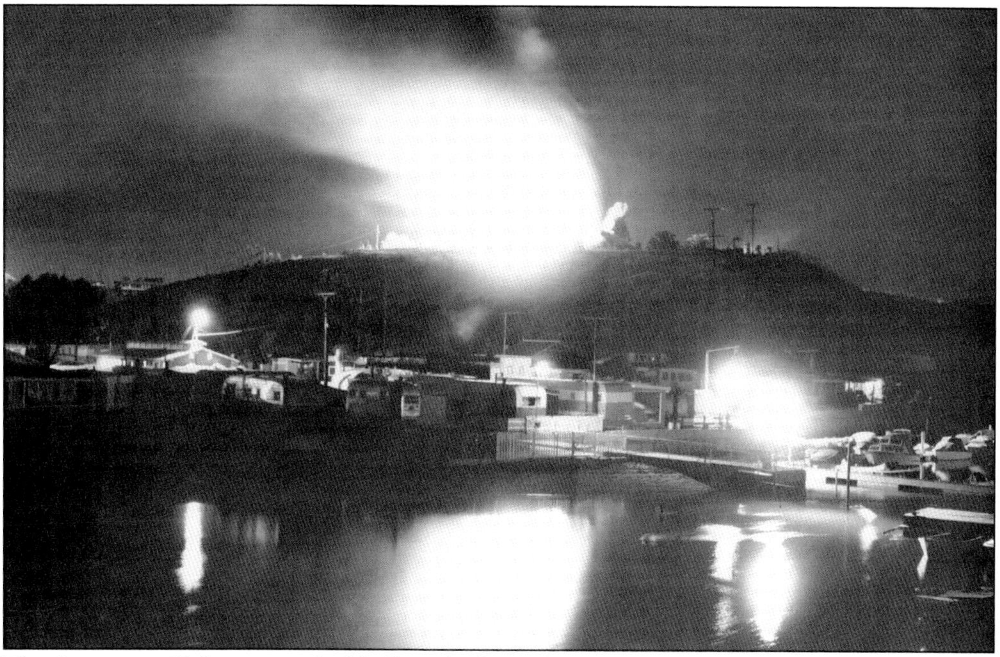

The Castaways Club sat atop the bluff overlooking the original Newport Landing. Built in 1904, the structure served as the Irvine Golf Course clubhouse and, before World War II, Mona's Studio Club. It was closed for security reasons during the war; the club's bright lights violated blackout restrictions. In November 1956, the popular restaurant and nightclub burst into a mass of flames that could be seen from Long Beach to San Clemente. (Courtesy of Richard Koehler.)

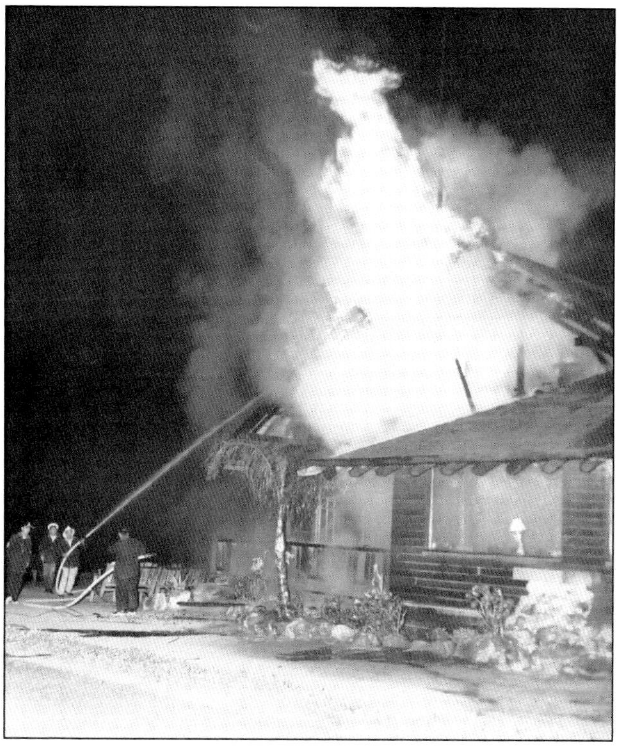

Show biz personality Frank Coburn owned the Castaways Club in 1956. He and his wife were awakened by smoke around 4:00 a.m. and four Newport Beach engines immediately responded. Firefighters were forced to lay hose hundreds of feet up a steep slope to reach the fire. The entire second floor and roof were fully involved by the time water was brought to bear on the flames. (Courtesy of Richard Koehler.)

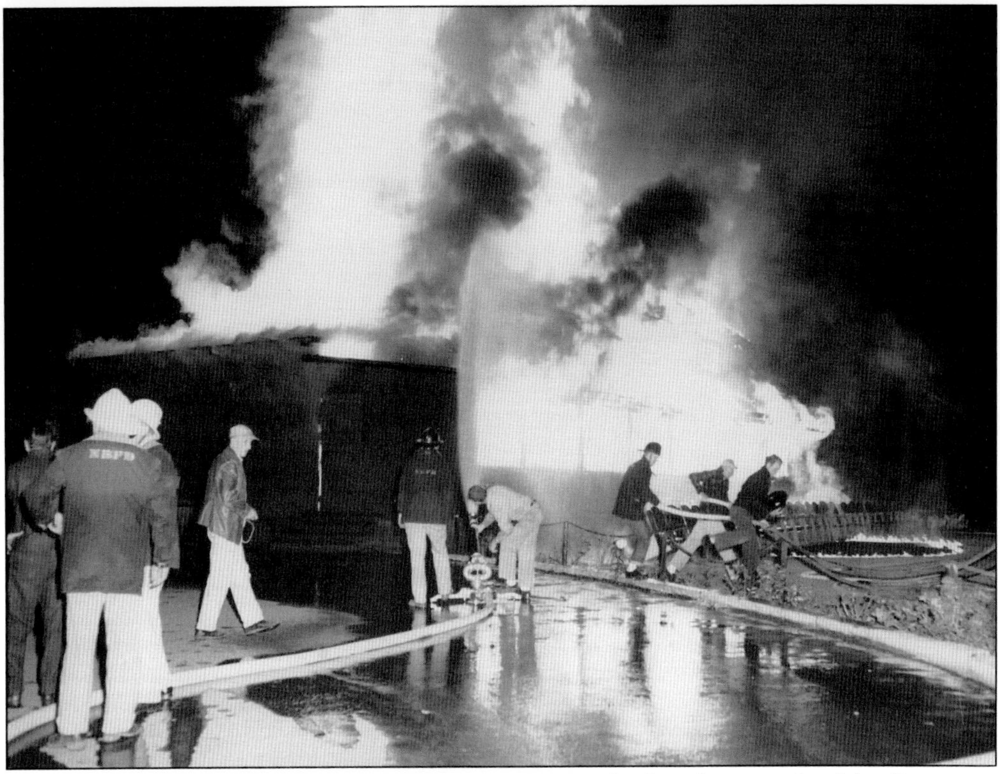

In this photograph of the Castaways fire, firefighters are setting up a monitor nozzle to apply a larger amount of water on the fire than is obtainable with hand-held lines. Other firefighters are still extending hose lines into position. The entire building is ablaze and beyond saving at this point. (Courtesy of Richard Koehler.)

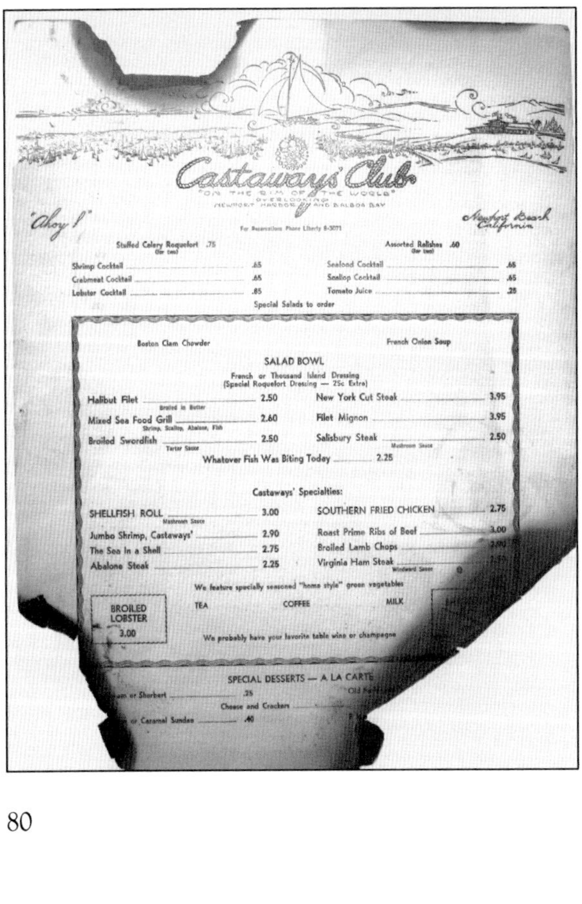

Castaways' Club
"ON THE RIM OF THE WORLD"
OVERLOOKING
NEWPORT HARBOR AND BALBOA BAY

"Ahoy!"

Newport Beach California

For Reservations Phone Liberty 8-3071

Stuffed Celery Roquefort .75 (for two)
Assorted Relishes .60 (for two)

Shrimp Cocktail .65
Crabmeat Cocktail .65
Lobster Cocktail .85
Seafood Cocktail .65
Scallop Cocktail .65
Tomato Juice .25

Special Salads to order

Boston Clam Chowder
French Onion Soup

SALAD BOWL

French or Thousand Island Dressing
(Special Roquefort Dressing — 25c Extra)

Halibut Filet (Broiled in Butter) 2.50
Mixed Sea Food Grill (Shrimp, Scallop, Abalone, Fish) 2.60
Broiled Swordfish (Tartar Sauce) 2.50
New York Cut Steak 3.95
Filet Mignon 3.95
Salisbury Steak (Mushroom Sauce) 2.50

Whatever Fish Was Biting Today 2.25

Castaways' Specialties:

SHELLFISH ROLL (Mushroom Sauce) 3.00
Jumbo Shrimp, Castaways' 2.90
The Sea in a Shell 2.75
Abalone Steak 2.25
SOUTHERN FRIED CHICKEN 2.75
Roast Prime Ribs of Beef 3.00
Broiled Lamb Chops [illegible]
Virginia Ham Steak (Windward Sauce) [illegible]

We feature specially seasoned "home style" green vegetables

BROILED LOBSTER 3.00

TEA COFFEE MILK

We probably have your favorite table wine or champagne

SPECIAL DESSERTS — A LA CARTE

...am or Sherbert .25
Cheese and Crackers
... or Caramel Sundae .40

The Castaways Club, renowned for nautical décor that included a large mural of a nude mermaid on horseback, had been reduced to what one fire official referred to as "marshmallow size." Only the stone chimney, foundation, and a few charred menus remained. The club was never rebuilt. Today the bluff is a public park. (Author's collection.)

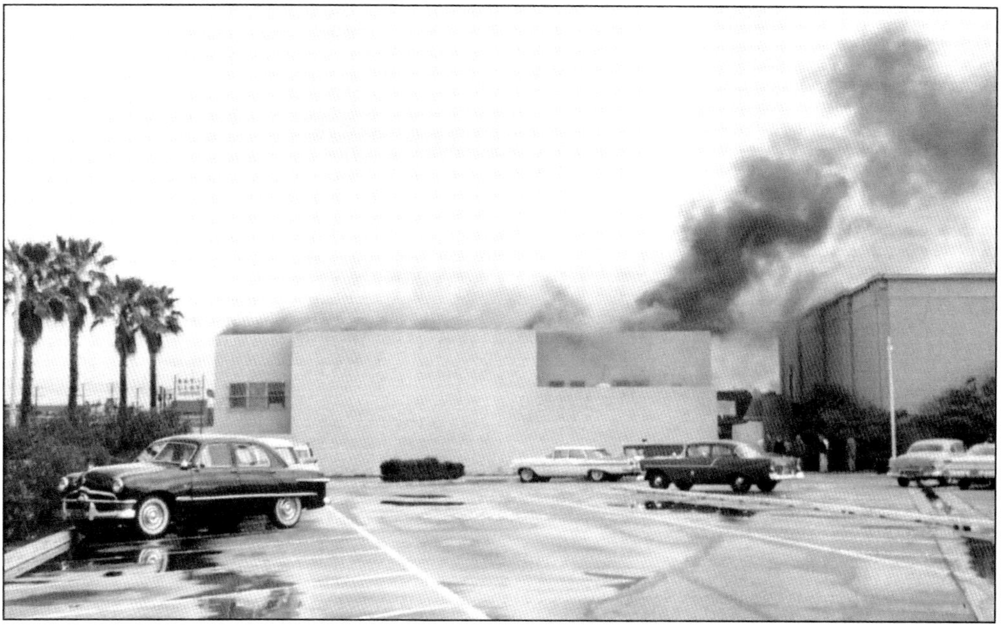

On March 28, 1960, Vincent's Lido Drugs, adjacent to city hall and the historic Lido Theatre, was struck by fire. A grease fire started in the kitchen and the cook was unable to control it with an extinguisher. The fire quickly traveled up a flue and spread throughout the attic. Dispatcher Hugh McMillan took this photograph as he ran over to the scene from Station Two. (Courtesy of NBFD archives.)

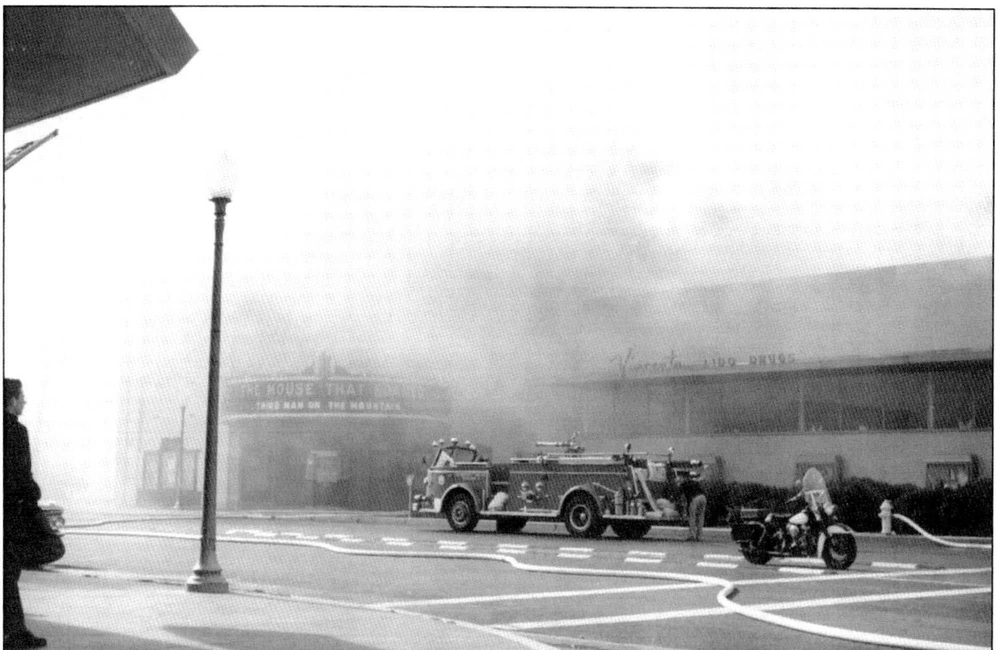

The fire was called in by a city employee from his office at nearby city hall. Two engines and a squad immediately responded. McMillan snapped this photograph of Engine No. 2 after it was connected to a hydrant and just before hose lines were extended. (Courtesy of NBFD archives.)

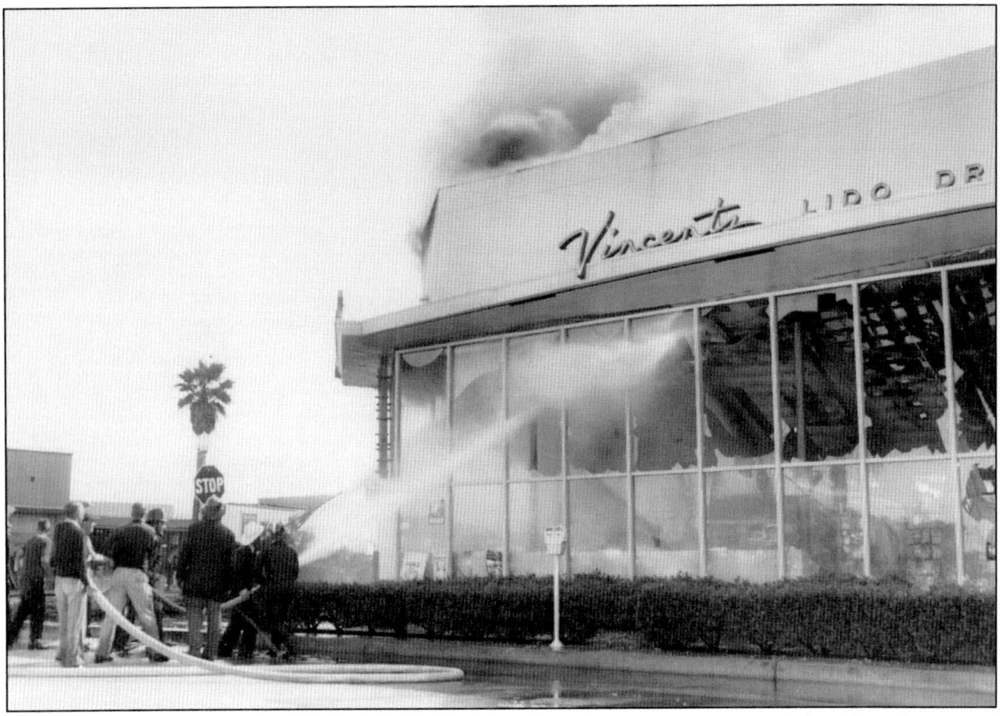

A large crowd gathered to watch. Several bystanders jumped in to lend a hand. Here they help firefighters direct a powerful stream of water into the interior of the building. (Courtesy of Richard Koehler.)

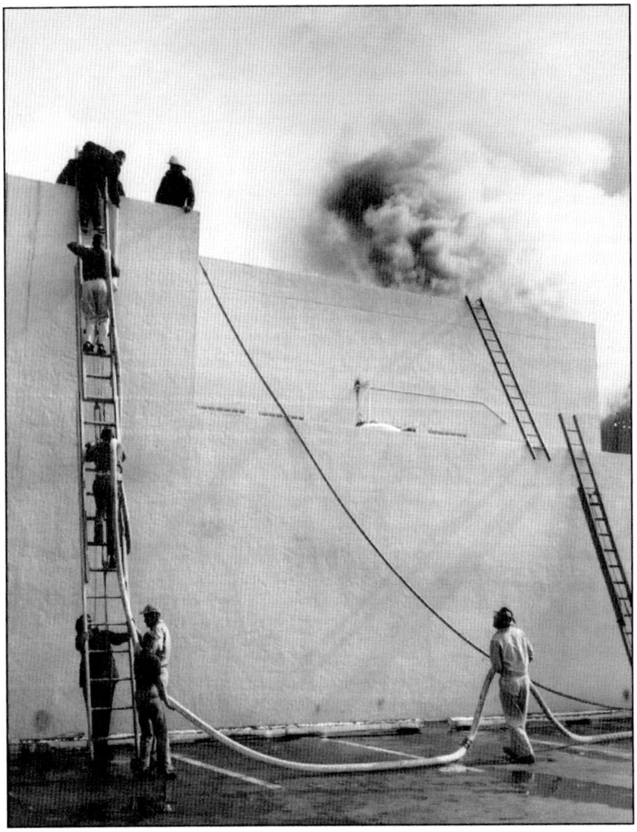

Assistant chief Phil Hayden, wearing a white helmet, directed firefighters on the roof. Shortly after this photograph was taken, Chief Hayden fell through the roof. He was able to arrest his fall, narrowly avoiding a tumble to the main floor. He injured his shoulder and back. Minutes later, he was up again and directing firefighting operations. (Courtesy of Richard Koehler.)

Firefighters abandoned the roof and the interior of the building. Shortly thereafter, the roof collapsed. Firefighters focused their efforts on protecting the nearby Lido Theatre from damage. (Courtesy of Richard Koehler.)

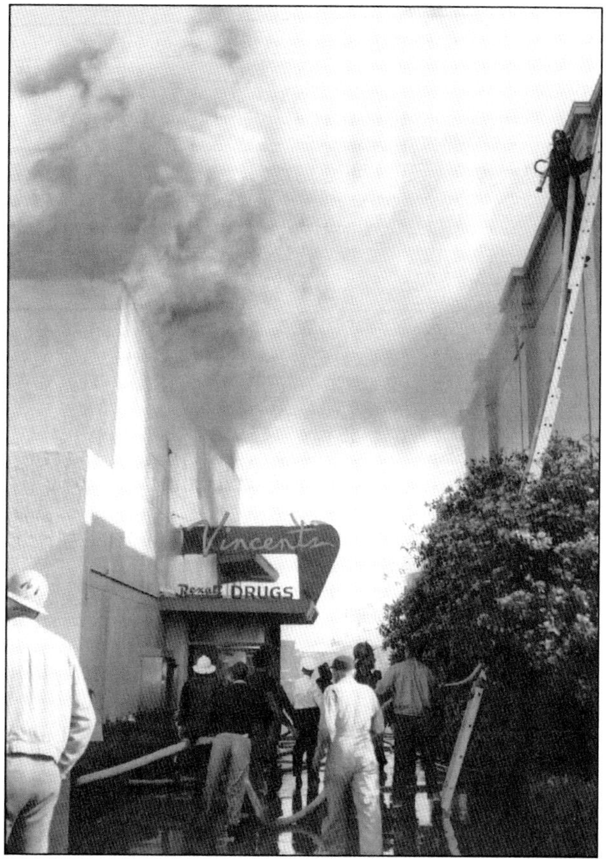

The front facade of the drugstore collapsed. All efforts to extinguish the fire were now being made from outside the building. In this photograph, Engine No. 2's deck gun has been brought to bear upon the interior. This large nozzle is capable of applying large quantities of water directly upon the fire. (Courtesy of Richard Koehler.)

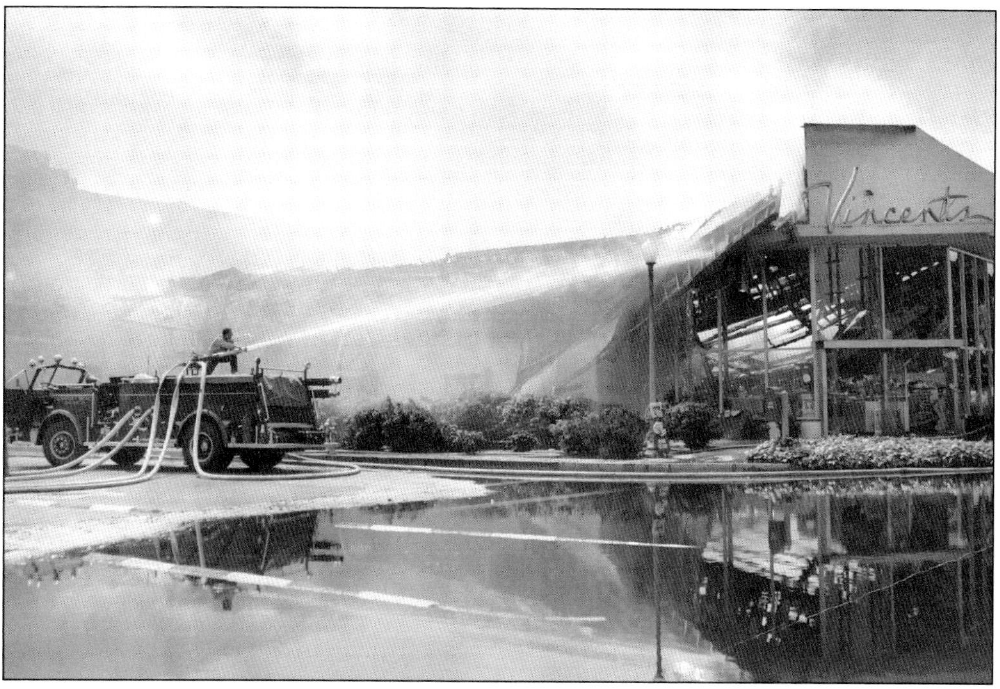

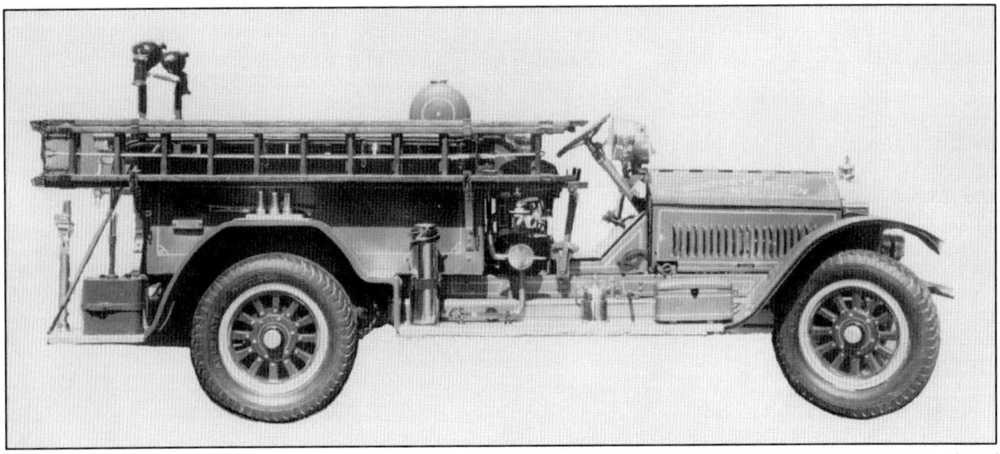

Chief Briscoe spent six years building this scale model of Newport's first fire engine. It is made of wood and exact in every detail. Today it sits in the fire chief's office in a glass case. While Briscoe was chief, pranksters would periodically sneak downstairs into his office with a hypodermic syringe and inject a few drops of motor oil onto the wood base beneath the model's motor. Try as he might, Chief Briscoe never caught the scofflaws in the act. (Courtesy of NBFD archives.)

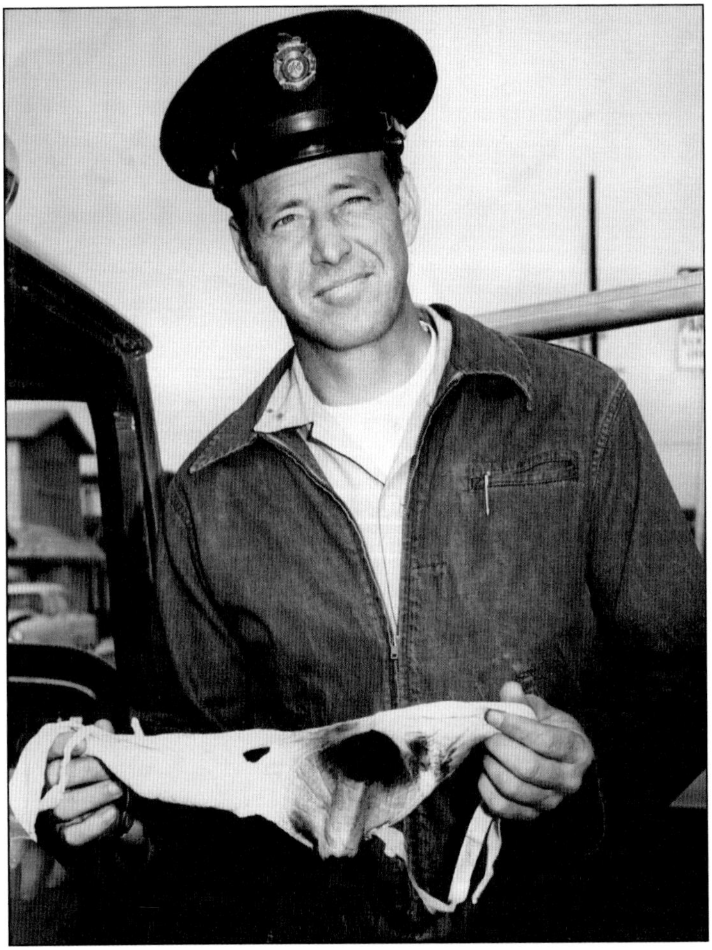

In March 1966, the *Daily Pilot* reported, "This Was No False Alarm: Where there's smoke, there is apt to be a fire, Newport Beach firemen discovered when they answered a 'false alarm' at a Laundromat at 200 30th Street. They discovered a smoldering brassiere on the Laundromat floor, apparently discarded after the rubber padding caught fire in a dryer. Firemen extinguished the garment and reported it a total loss." Ted McMasters displays the culprit. (Courtesy of NBFD archives.)

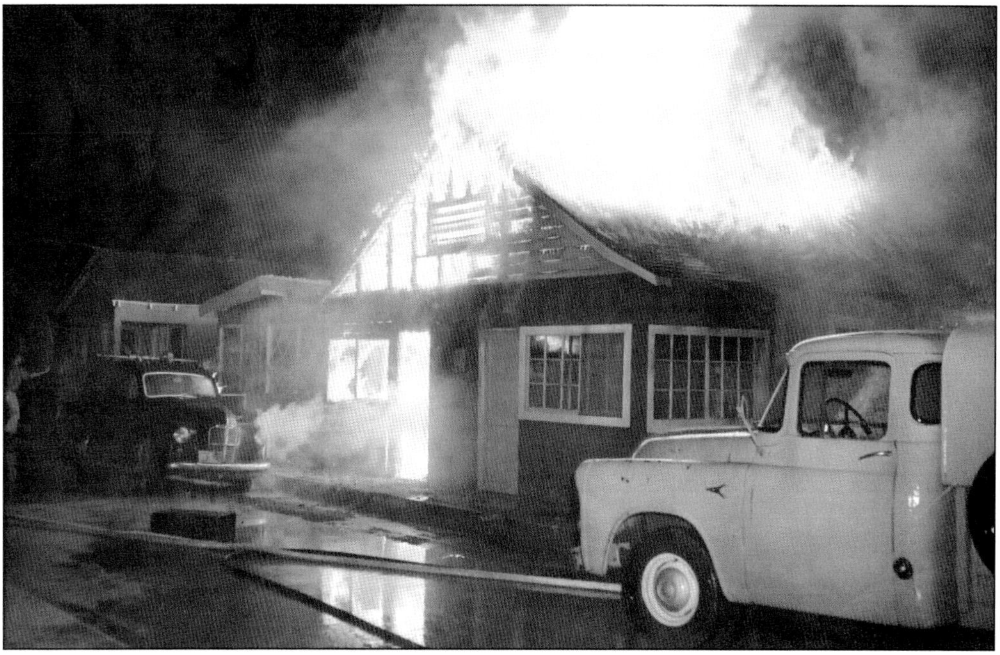

Two men went out on the town one night in 1960, leaving their buddy asleep in this house. They returned briefly at one point to find their friend still sleeping and a cigarette smoldering in his mattress. They threw some beer on the small fire and left the house. Firefighters were called to the scene shortly thereafter, finding the house completely engulfed in flames and a man who died without waking inside. (Courtesy of Richard Koehler.)

FIREMAN

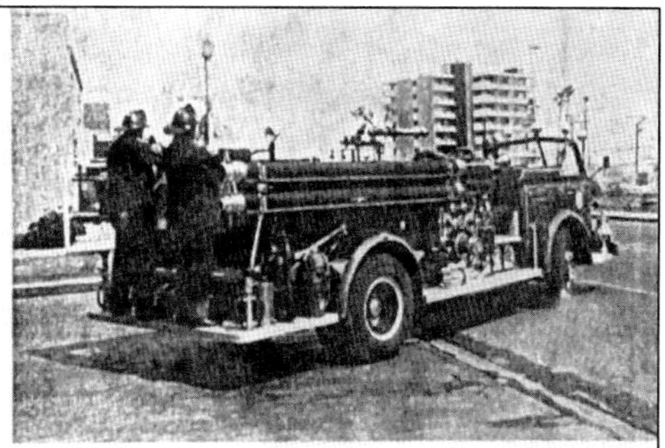

CITY OF NEWPORT BEACH
OPEN COMPETITIVE EXAMINATION

SALARY: $510-$620 per month (paid biweekly). First merit increase to $536 after six months. Yearly merit increases thereafter.

FILING DATE: Applicants may file through Friday, November 6, 1964 at 5:00 P.M.

TEST DATE: The written, physical agility and swim tests will be given on Saturday, November 7, 1964, 8:30 A.M. at Costa Mesa High School. Those successful will be asked to appear for an oral interview at a later date.

Paid-call firefighters were phased out in the early 1960s. By 1972, three platoons of paid firefighters manned the stations. The author remembers sleeping overnight on the sidewalk outside the Newport Beach City Hall during the late 1970s to be sure of receiving an application for the coveted position of firefighter. Today the process is a bit more civilized, but the position is no less desirable. (Courtesy of NBFD archives.)

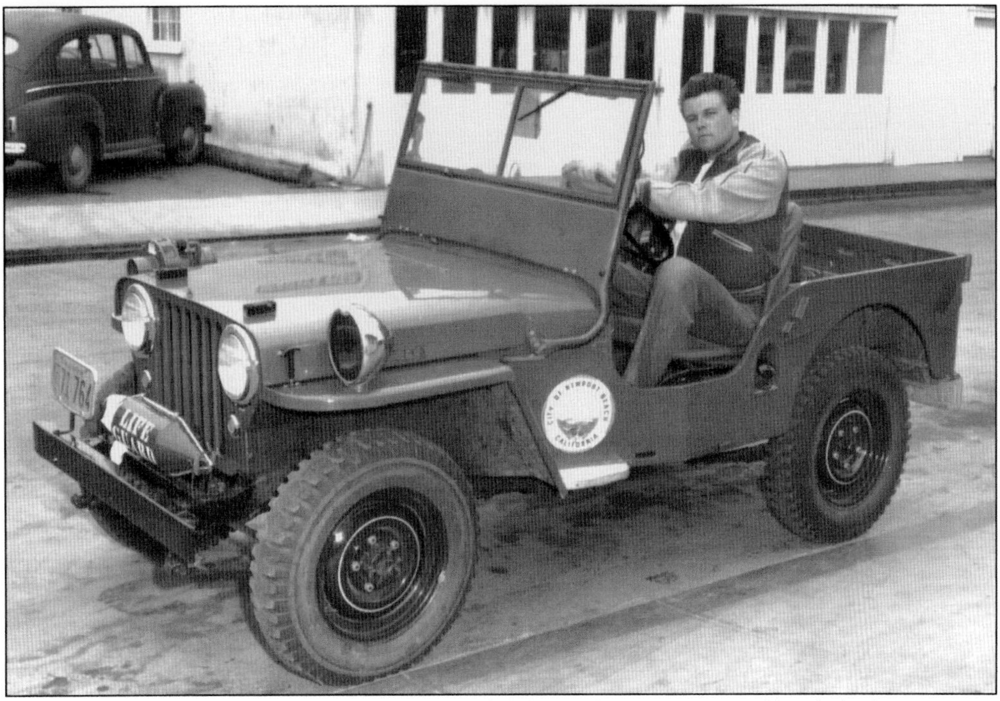

Jack Jones started his career as a Newport Beach lifeguard in 1942. One of his daily duties was to lower a glass bottle from the end of the Newport Pier to collect a sample of seawater to record its temperature. When the lifeguards separated from the fire department, Jones cast his lot in with the firefighters. (Courtesy of NBFD archives.)

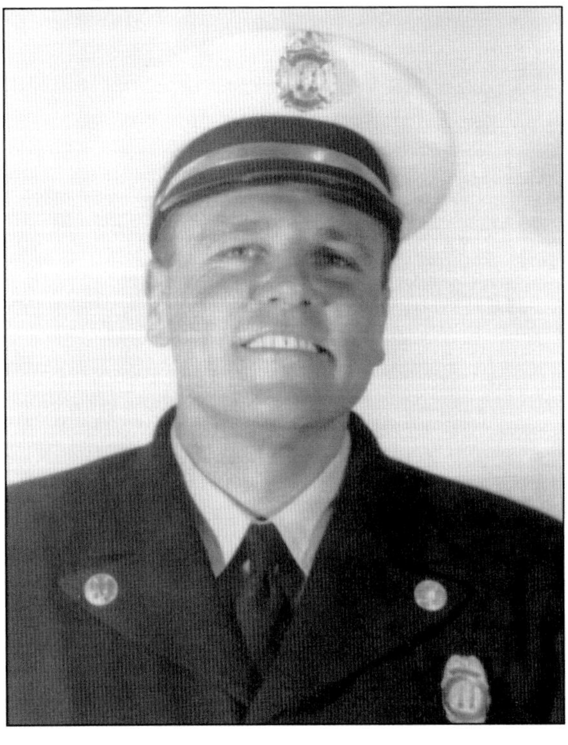

In March 1964, Santa Ana winds knocked down a power line that quickly ignited a grass fire. A three-man engine company was dispatched. Capt. Jim Brooks started to climb over a barbed-wire fence with a hose line. In the dark, no one noticed the power line lying across the fence. Brooks was immediately knocked unconscious by 4,000 volts of electricity. Firefighter William "Dutch" Van Horn tried to pull Brooks off the fence. He too was knocked to the ground. Without hesitation, Jack Jones grabbed Brooks and pulled him off of the fence. Engineer Jones began mouth-to-mouth resuscitation until help arrived. Brooks and Van Horn were taken to the hospital where both recovered. (Courtesy of NBFD archives.)

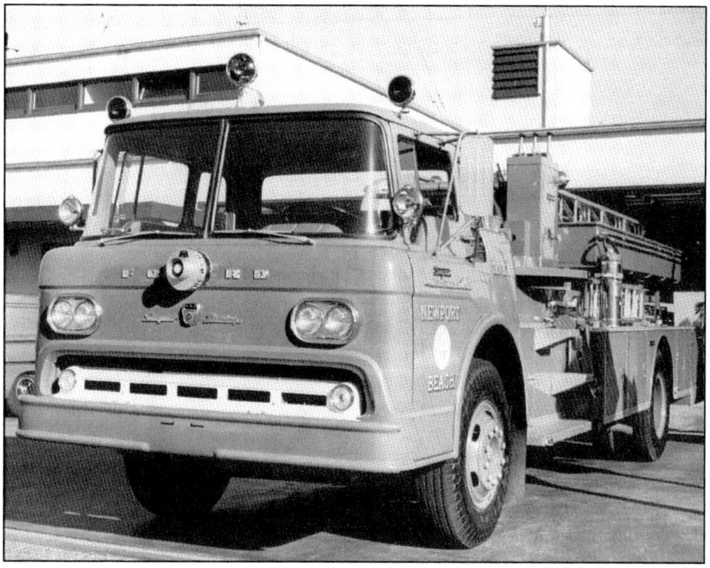

The NBFD purchased its first aerial ladder truck, a Ford/Seagraves combination, in 1960. Truck No. 1 carried an 85-foot ladder and an assortment of tools, including a life net. Four firefighters were assigned to this truck for a brief period. There was not enough room in the cab for all, so the least senior firefighter donned a helmet and safety belt and braved the elements atop the ladder turntable. (Courtesy of Richard Koehler.)

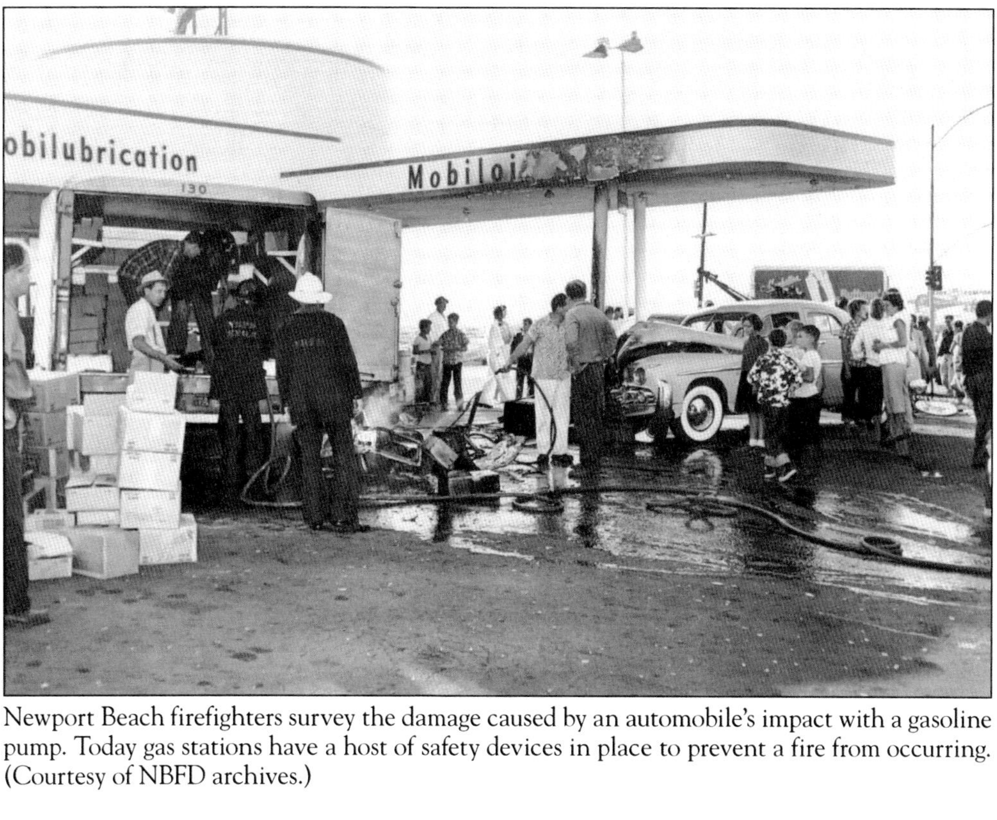

Newport Beach firefighters survey the damage caused by an automobile's impact with a gasoline pump. Today gas stations have a host of safety devices in place to prevent a fire from occurring. (Courtesy of NBFD archives.)

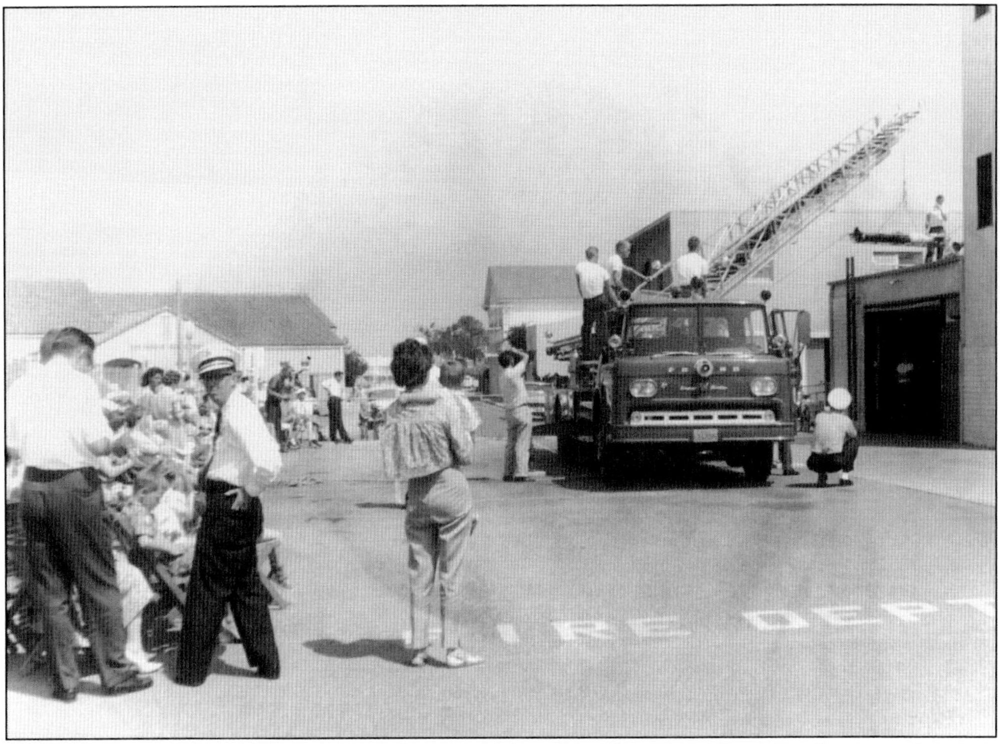

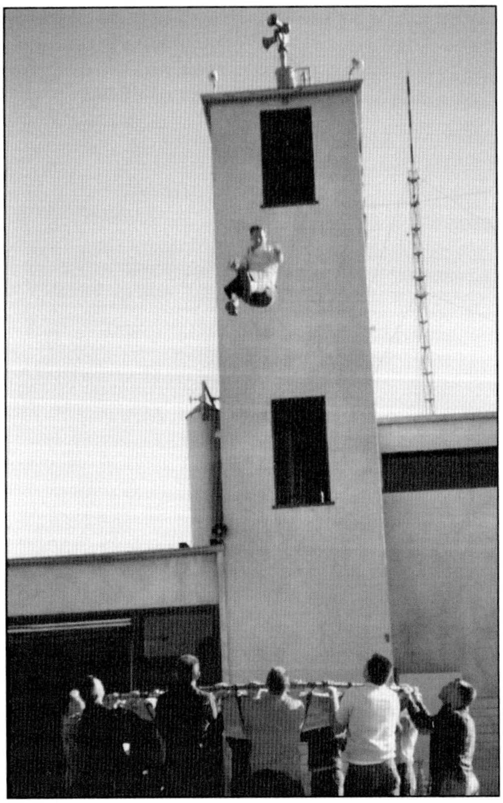

In March, 1932, Chief Crocker issued an invitation to Newport Beach residents to visit their local fire stations to talk to firefighters about their jobs. Chief Crocker believed the general public was convinced that firefighters spent their day playing checkers—an open house would be a good way to dispel this notion. Open house soon became an annual event. In this photograph, the public is watching a demonstration of truck company operations behind Fire Station Two. (Courtesy of NBFD archives.)

In this photograph, firefighters demonstrate the use of a life net. Life nets were carried on Newport Beach trucks until the end of the 1970s when they were deemed to be unsafe. In its day, this rarely used piece of equipment was a definite crowd pleaser at open house. (Courtesy of NBFD archives.)

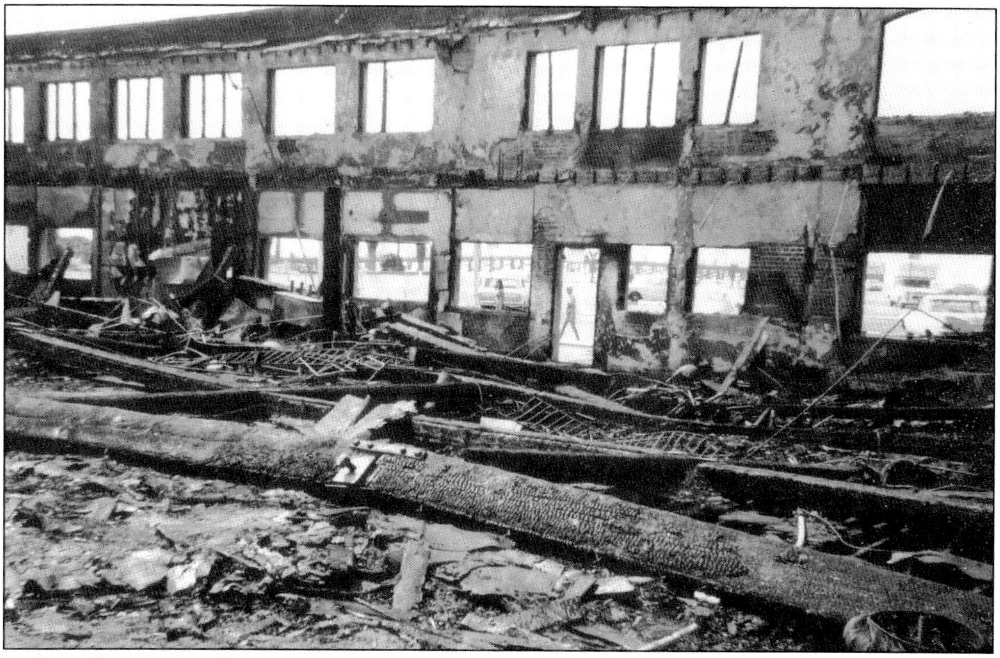

Early in the morning hours of August 8, 1966, fire struck the Rendezvous Ballroom for the second time. Once again, a discarded cigarette was the culprit. Seventeen firefighters and seven engines fought the fire for an hour but the Rendezvous was beyond saving. The famous ballroom would not rise from the ashes a second time. (Courtesy of NBFD archives.)

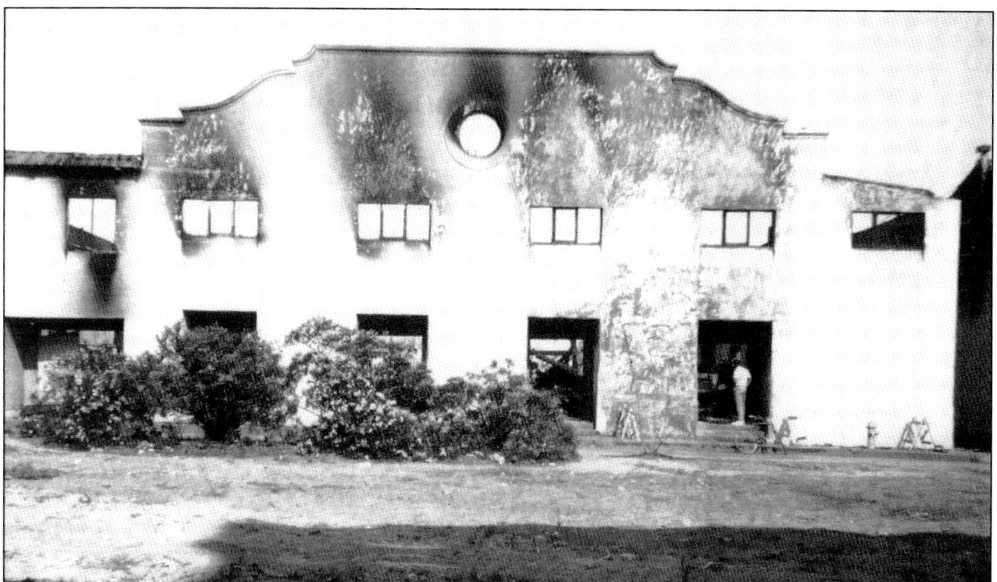

The Rendezvous had prospered after the 1935 fire, hosting the Big Band sound of Stan Kenton and others during the 1940s. The rock 'n' roll of Dick Dale and the Deltones and others emanated from the ballroom during the 1950s and 1960s. Ironically, a rock group called The Cindermen was the last to play from the Rendezvous stage. (Courtesy of NBFD archives.)

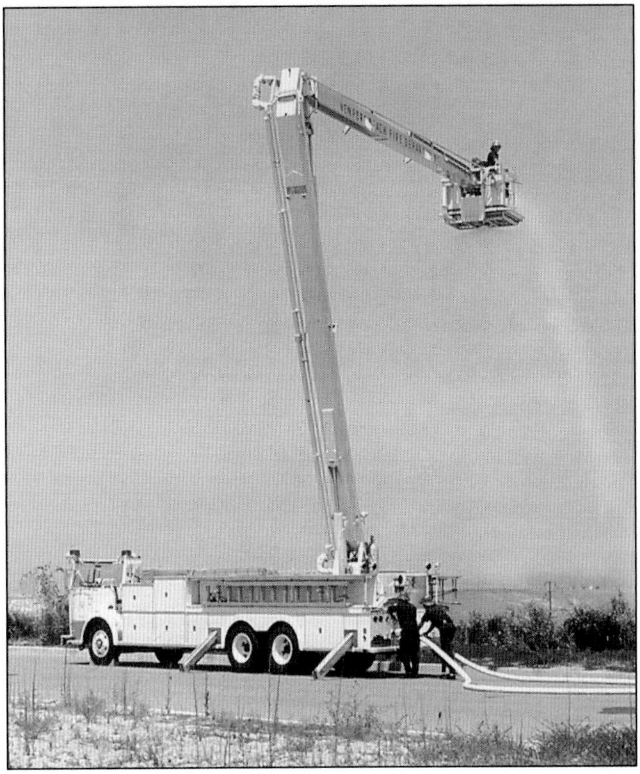

A Crown Snorkel was purchased in 1970. Crown was a well-respected manufacturer of fire apparatus. Newport's Crown featured a 75-foot articulated boom built by Snorkel. The boom concept was a departure from the traditional aerial ladder. (Courtesy of Bob Kelly.)

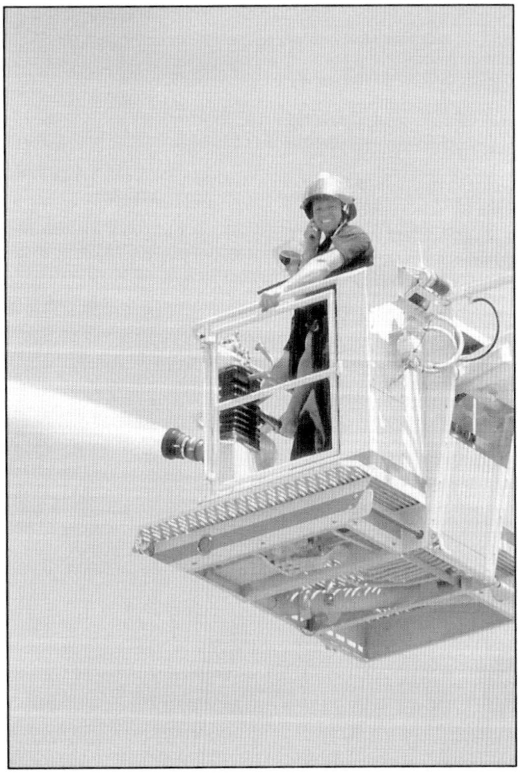

Capt. Jerry Strom operates the Snorkel boom and master stream nozzle 75 feet in the air. An additional five feet of elevation could be obtained by using a non-authorized procedure. Lowering the basket using this method often resulted in a few feet of free fall—and some white-knuckle grasping of the basket railing. (Courtesy of Bob Kelly.)

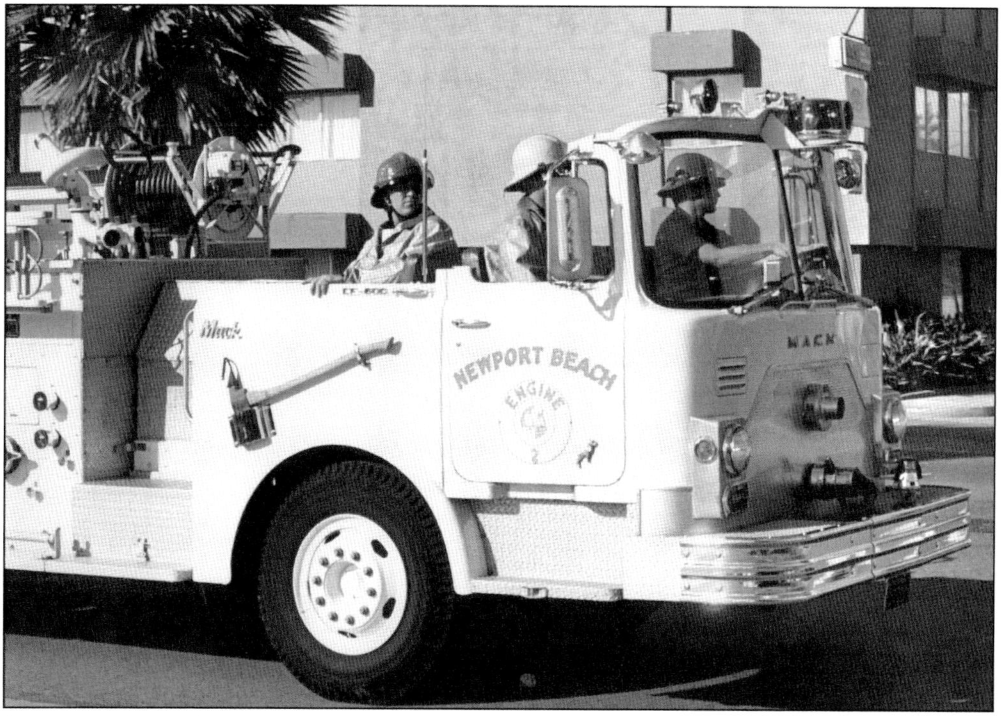

Several Mack CF Series fire engines were purchased over the years. All of Newport's engines and trucks purchased during the 1960s and 1970s were painted white. Because of the temperate climate, many had their roofs removed. By 1980, this practice was discontinued for safety reasons. (Courtesy of NBFD archives.)

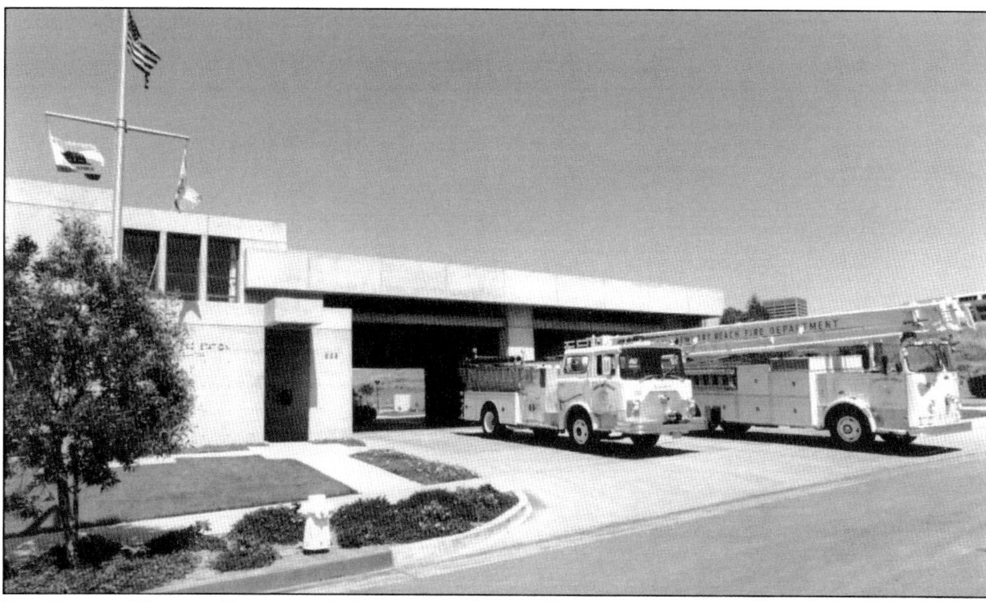

Fire Station Three, affectionately known as The Rock, opened in 1972 a mile away from Balboa Island. Budget-conscious council members suggested that Balboa Island could be covered by fire units from this new station, allowing Station Four to be closed. Balboa Island residents, intent on saving their fire station, flexed their political muscle and prevailed. (Courtesy of NBFD archives.)

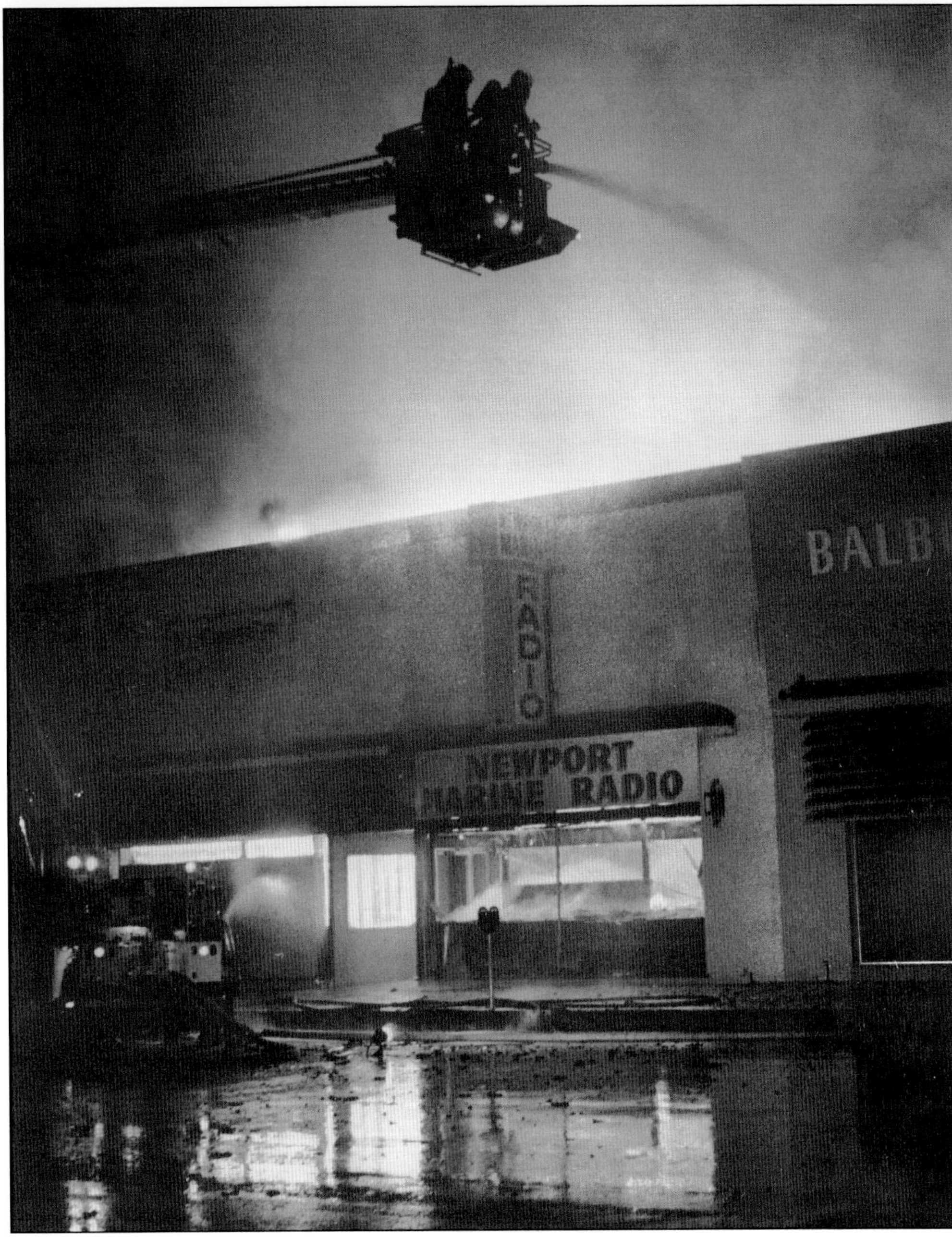

The worst conflagration in the 100-year history of the Newport Beach Fire Department occurred on January 11, 1975. Police officer Michael Jackson was on patrol around 2:30 a.m. when he noticed smoke coming from a block-long stretch of commercial storefront along Pacific Coast Highway known as Mariner's Mile. These businesses shared a common attic, with no fire stops between the occupancies. Fire spread from the office of the local cable television provider into the attic and quickly traveled in both directions. Fire department units immediately responded, only to find the entire block on fire. Fire chief Leo Love was called at home, and he later stated, "When I left my home in Costa Mesa, it looked like the whole world was on fire." (Courtesy of Richard Koehler.)

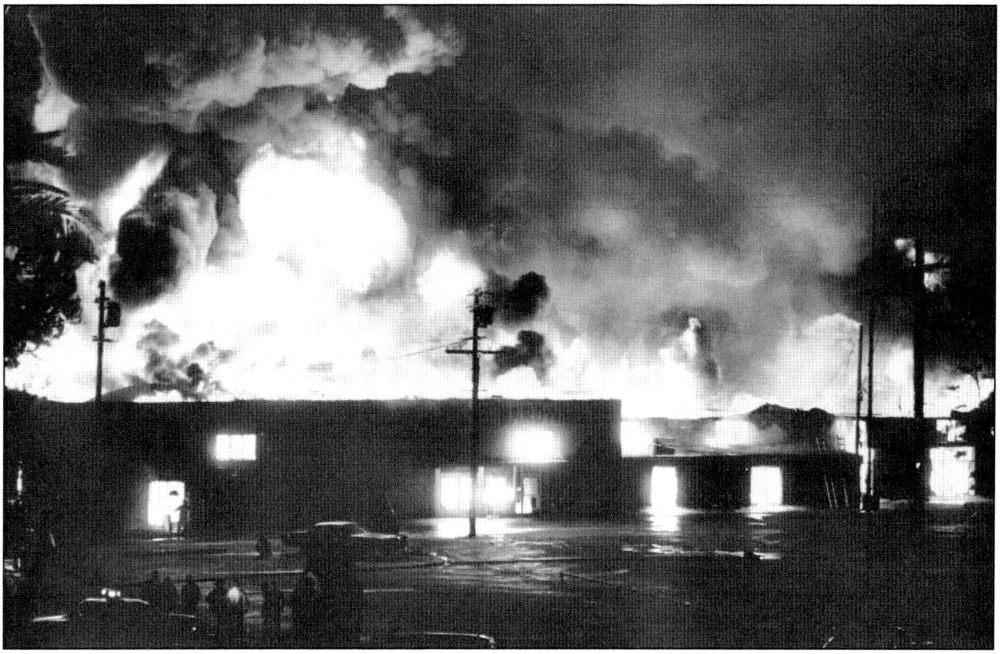

Firefighters quickly discovered that the water pressure in nearby fire hydrants was inadequate. They were forced to stretch hose over great distances to obtain a water supply. Explosions, fueled by paint and aerosol cans, shattered the night as the fire intensified. Surplus signal flares shot through the flames above the marine supply store. Firefighters struggled to put out spot fires as sparks drifted across Pacific Coast Highway. (Courtesy of Richard Koehler.)

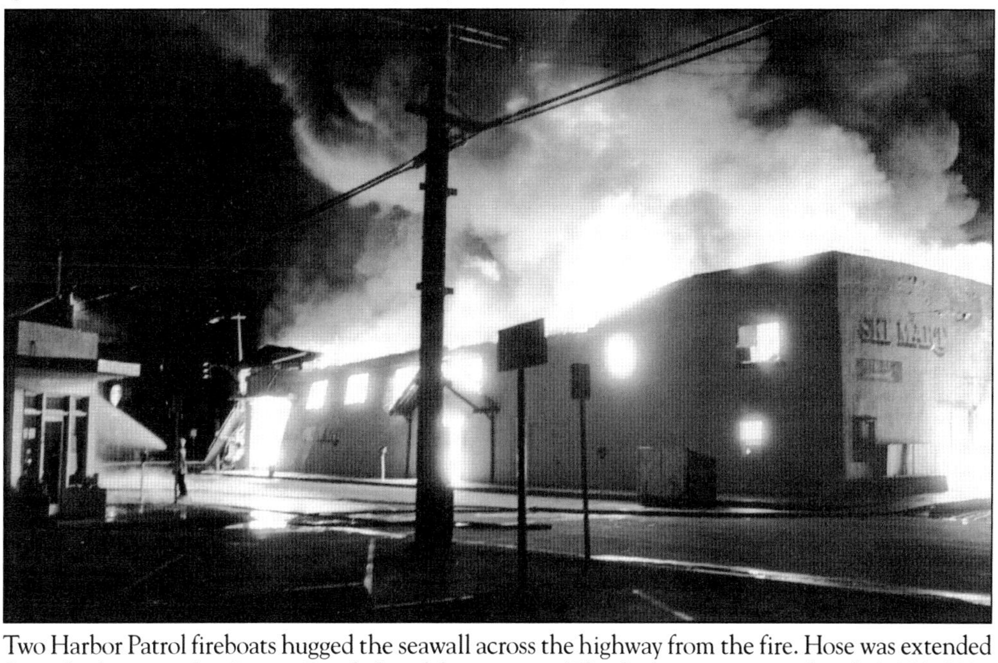

Two Harbor Patrol fireboats hugged the seawall across the highway from the fire. Hose was extended from the boats to the Crown Snorkel and fire engines. The fireboats, equipped with 2,000 GPM fire pumps, were able to provide much needed water to the fire scene. It was too late. Firefighters watched helplessly as wall after wall collapsed into the inferno. (Courtesy of Richard Koehler.)

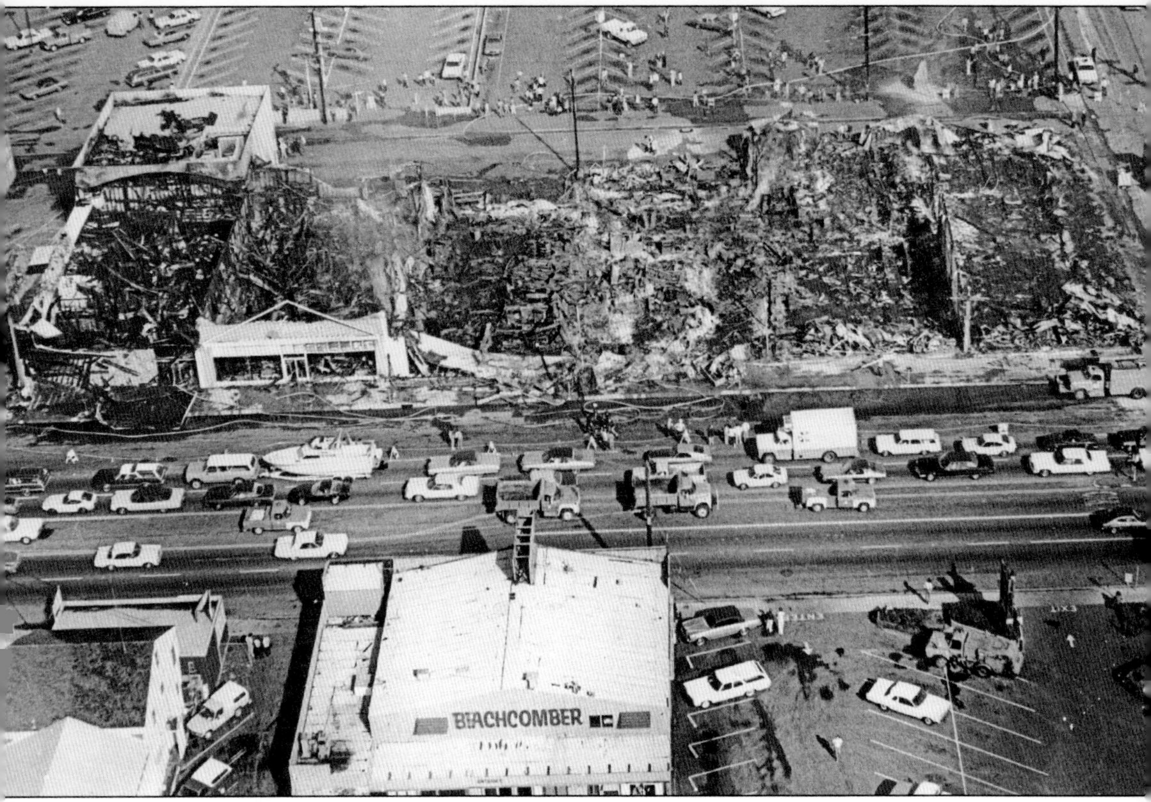

Newport firefighters fought the Mariner's Mile fire into Saturday night. A small flare up around 10:30 p.m. was quickly extinguished. This photograph illustrates the devastation visible on Sunday. Chief Love summed up the fire, "Too much fire, too fast." Investigators spent days sifting through the rubble searching for the cause of the fire. In the Ski Mart, cash registers and fiberglass skis were melted into unrecognizable lumps. Employees of the United California Bank found the bank vault to be damaged but intact. Firefighters were on scene in three minutes and their valiant efforts were beyond reproach. The Mariner's Mile fire brought about a much needed review of Newport Beach water distribution systems, and improvements would soon be made. (Courtesy of NBFD archives.)

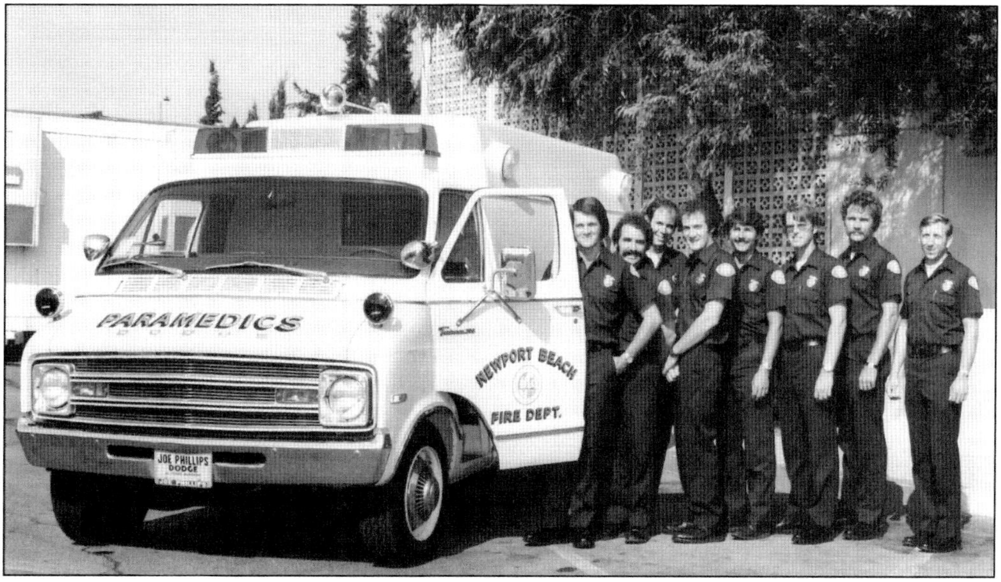

Fire departments have provided basic first aid for years. In 1970, Seattle launched the first paramedic unit, consisting of firefighters certified under an intensive advanced life-support training program. Around the same time, paramedics and firefighters were popularized across the country with the premiere of a new television show called *Emergency*. Early in 1975, eight Newport Beach firefighters graduated as certified paramedics. (Courtesy of NBFD archives.)

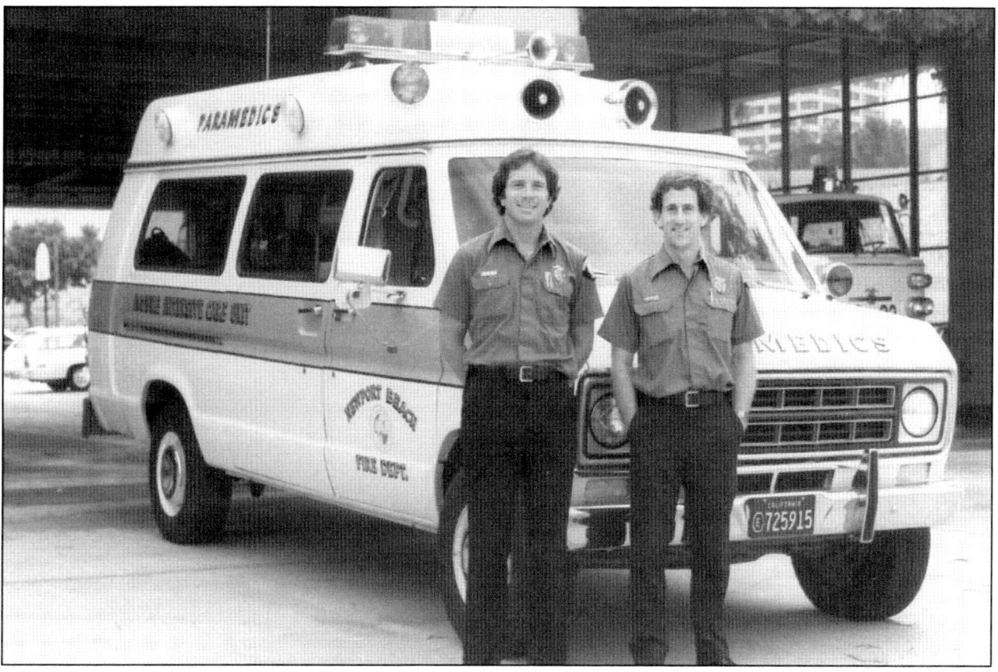

The advent of the paramedic program and the mandated certification of all fire personnel as emergency medical technicians signaled a shift in emphasis toward emergency medical response during the 1970s. This new direction didn't sit well with some Newport Beach old-timers and many chose to retire. As they left, a new generation of multifaceted firefighters took their place. (Courtesy of NBFD archives.)

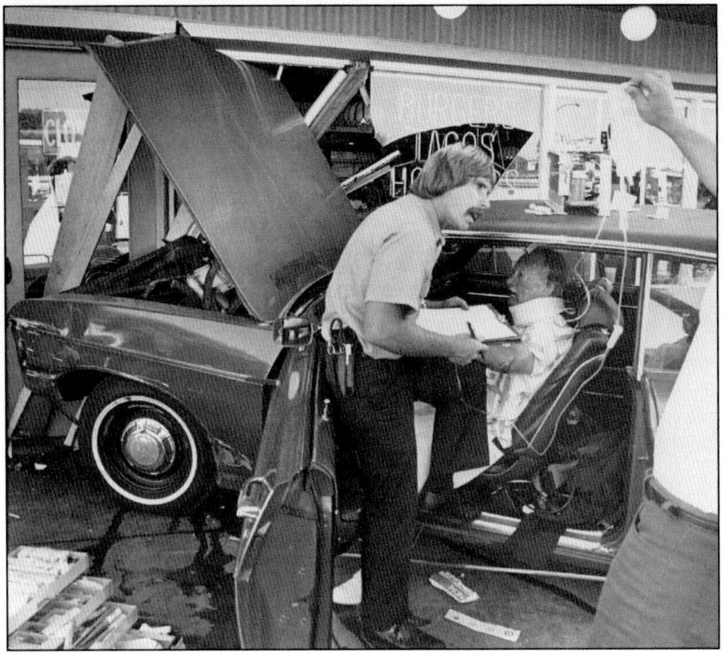

Rapid delivery of emergency care is critical to patient survival. The paramedic van carries everything needed to provide advanced life support at the scene of a medical emergency. Paramedics can identify problems, initiate a proper course of treatment, and stabilize the patient for transport. One medic accompanies the patient to the nearest hospital in a private ambulance, closely monitoring his or her condition. (Courtesy of NBFD archives.)

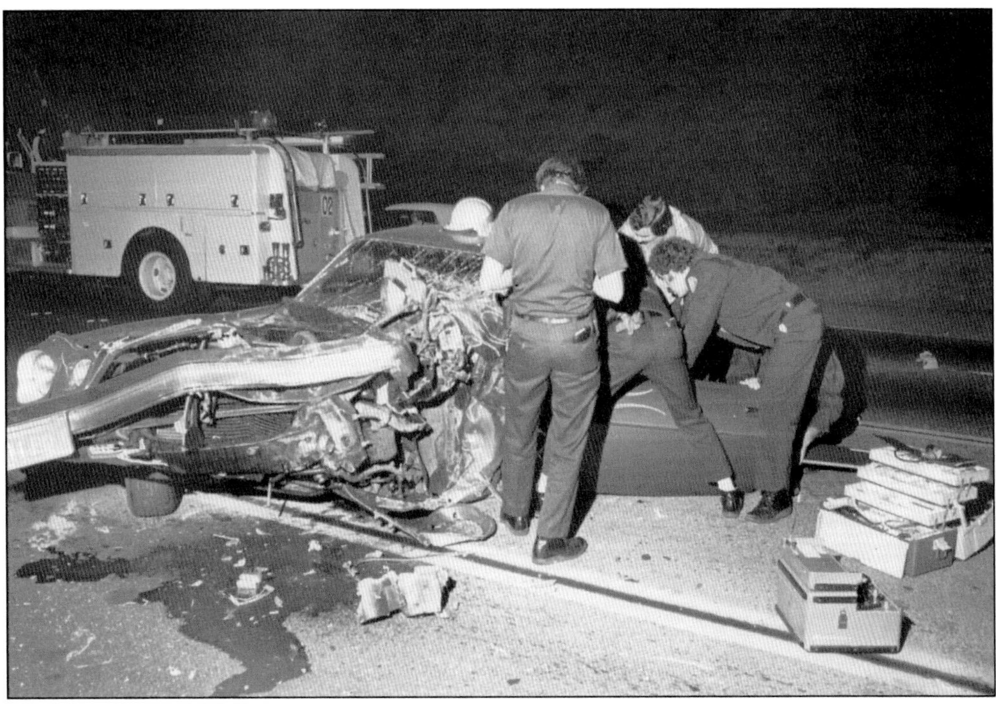

In 1975, a single paramedic van covered the entire city of Newport Beach. An engine company was dispatched to all medical emergencies, along with the paramedic unit. This insured that citizens would receive timely medical aid wherever they lived, along with specially trained care providers. Five firefighters would always be on hand to deal with any complications that might arise. (Courtesy of Richard Koehler.)

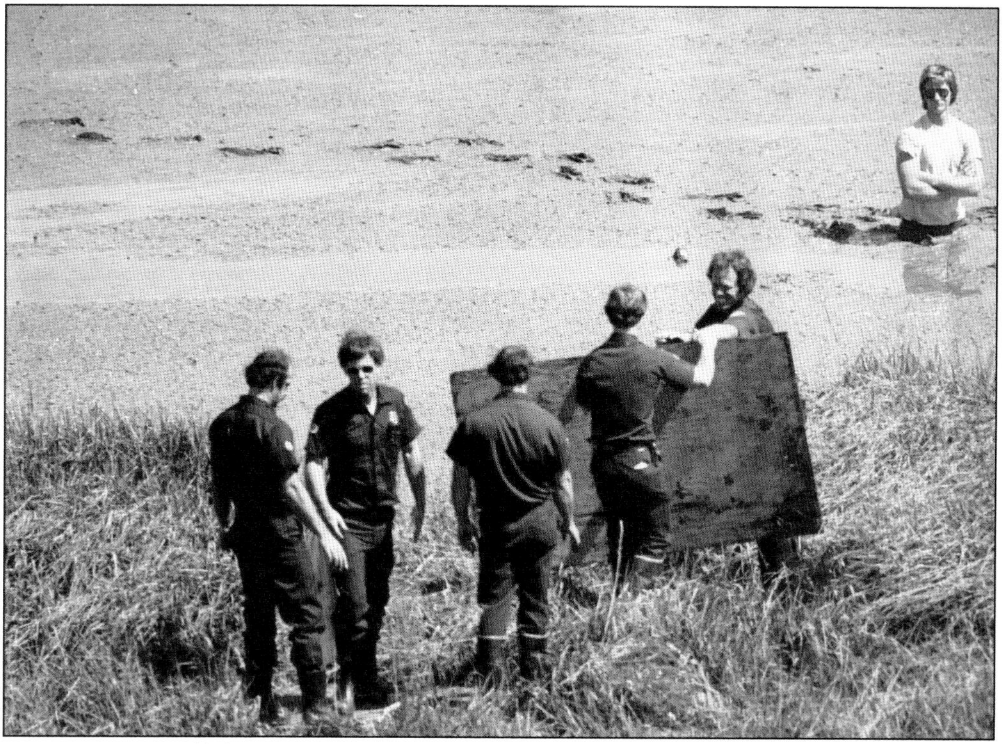

Rescues come in all shapes and sizes. A huge estuary and wildlife sanctuary lies inland of Newport Harbor. At low tide, the firmness of the exposed mud can be deceiving. This jogger tried to take a shortcut, sank to his waist, and found himself trapped. Firefighters utilized sheets of plywood to reach him. With a little digging, he was freed and on his way. (Courtesy of Richard Koehler.)

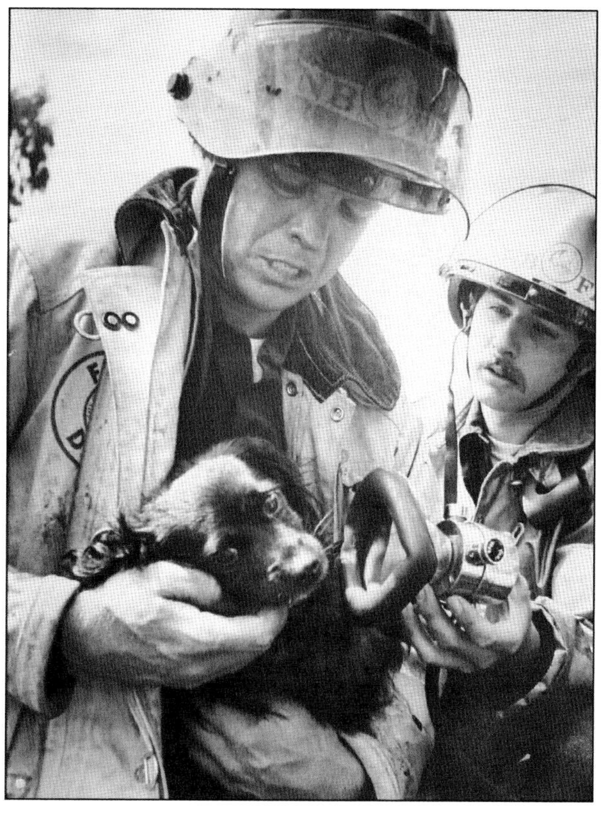

Capt. Jerry Strom and firefighter Ken Flessel administer oxygen to Puff the dog. Firefighters once climbed a tree to retrieve a child stuck in a tree house, but rarely did they ascend the heights to rescue a cat. Cats can be counted on to return to the ground at their own volition. (Photograph by Patrick O'Donnell; courtesy of Jerry Strom.)

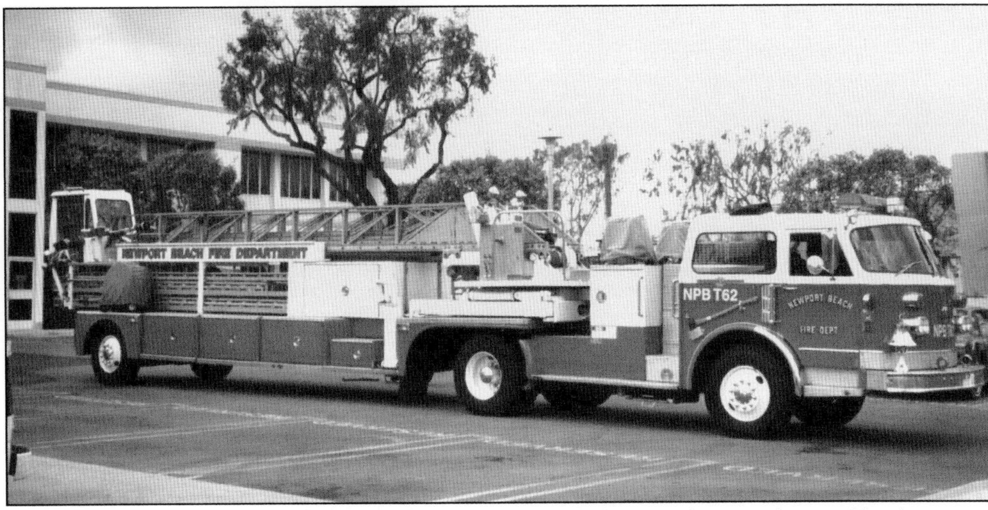

In 1978, Newport took delivery of its first tillered aerial ladder truck, manufactured by American LaFrance. The author remembers that one of his first assignments as a rookie firefighter was to get trained as a driver in the tiller "bucket" so that this truck could be put into service. (Courtesy of NBFD archives.)

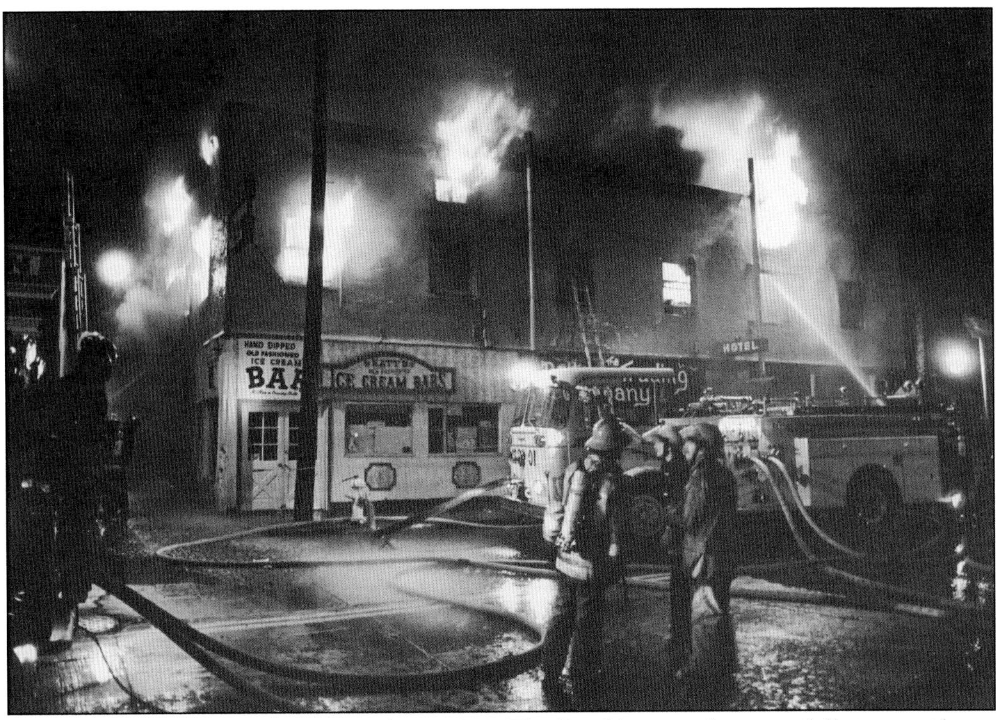

The Bay View Hotel went up in smoke in 1982. The Bay View was home to full-time residents, and investigators determined that a clandestine chemical lab started the fire. The Bay View was next to the Balboa Fun Zone and the Balboa Island Ferry, just a stone's throw from Newport's original Fire Station One. It took firefighters and fireboats three and a half hours to knock down the fire. (Courtesy of Richard Koehler.)

This fire, in a commercial area near Orange County's John Wayne Airport, was confined to one building. Battalion chief Augie Wagner described the blaze, "It just exploded. The fire was through the roof when we arrived." These elevated streams provide an effective method of extinguishing fire once the roof has collapsed and thousands of gallons of water are necessary to cool the interior. (Courtesy of Richard Koehler.)

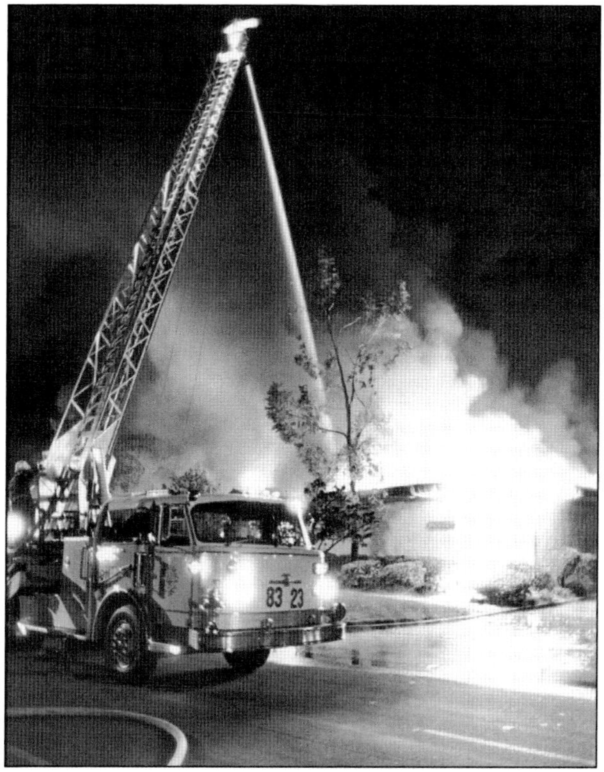

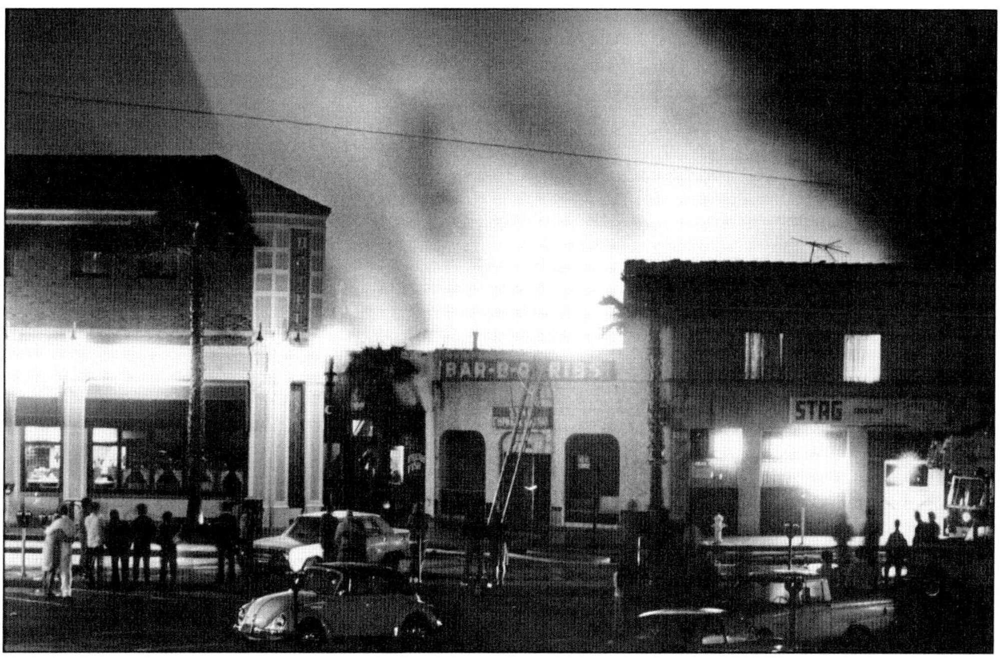

The Blue Beet was a popular bar and nightclub located next to the Newport Pier. The building, dating to 1912, was once occupied by an establishment reputed to be one of the last places to abolish slot machines in California. In April 1986, the Blue Beet went up in flames. (Courtesy of Richard Koehler.)

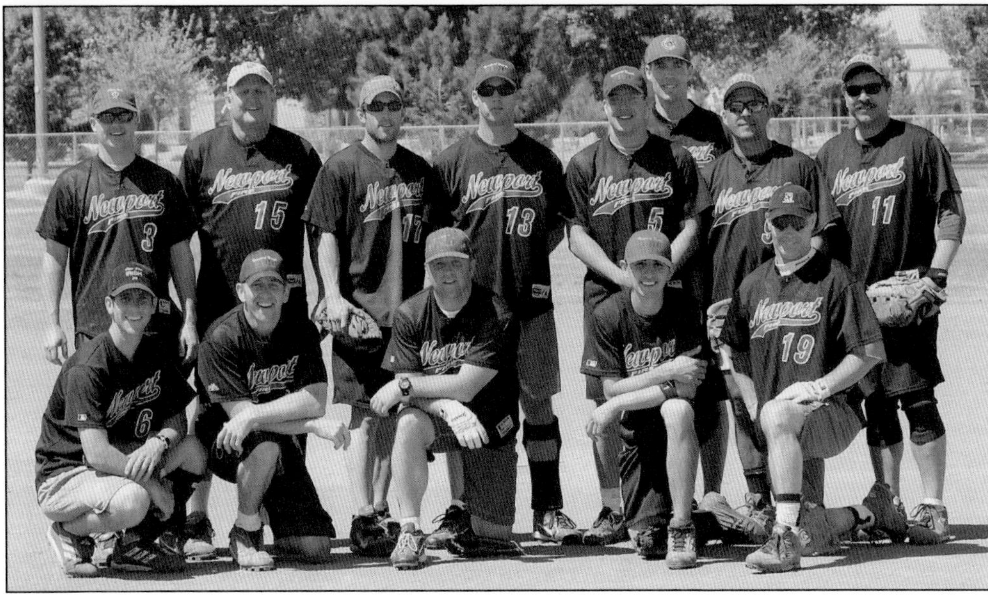

The men and women of the Newport Beach Fire Department have a long tradition of representing the department in spirited athletic competition. Organizing a softball team was first mentioned in 1916. A competitive team has been fielded virtually every year since. (Courtesy of Jeff Remmele.)

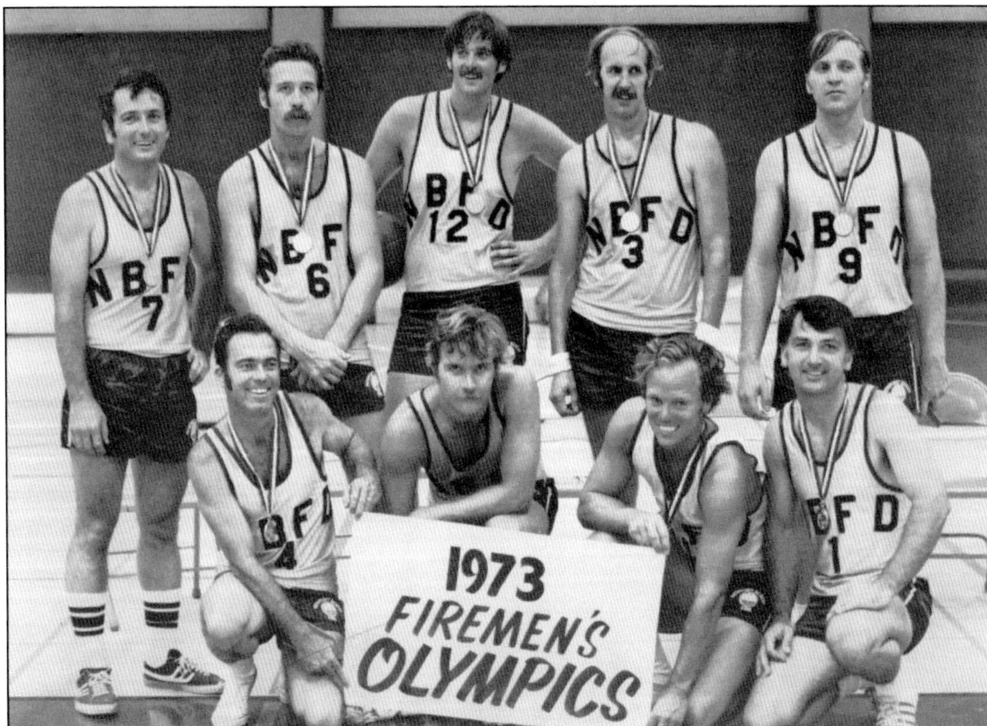

Newport Fire basketball teams have done particularly well over the years, especially in the annual California Firefighter's Summer Olympics. Jim Reed, no. 7, was one of three Newport fire chiefs with exceptional basketball skills. Members have also medaled in skiing, surfing, paintball, tennis, soccer, and golf. (Courtesy of NBFD archives.)

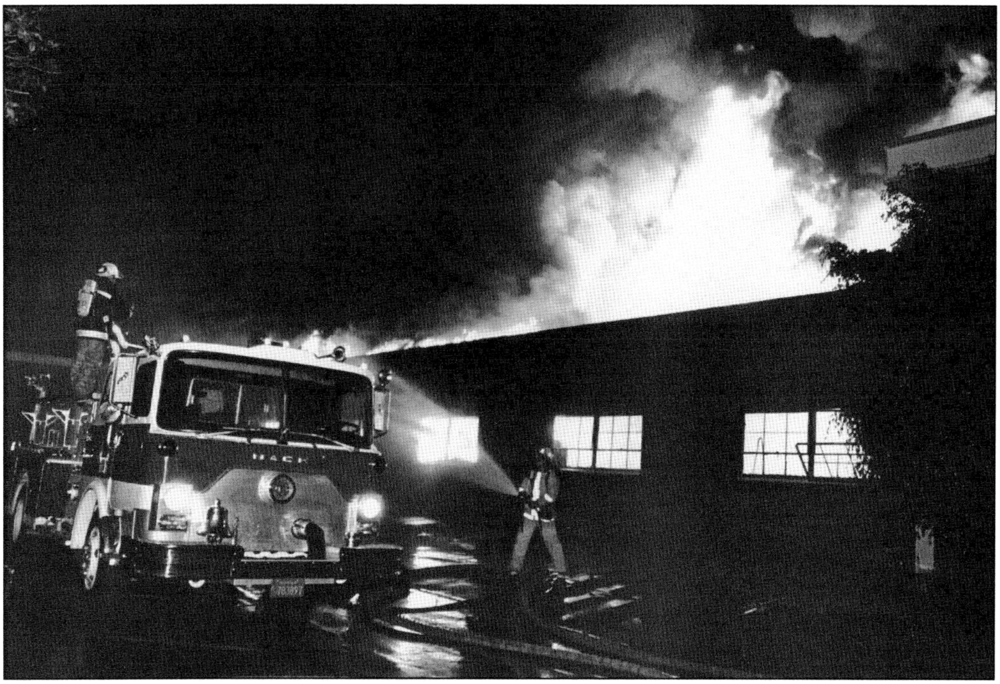

Fire struck two adjacent offices of the National Education Corporation near John Wayne Airport in 1983. The first, in July, was spotted by the police helicopter. In this photograph, a firefighter can be seen hosing down an engine after its windshield cracked and emergency warning lights melted in the intense heat. (Courtesy of Richard Koehler.)

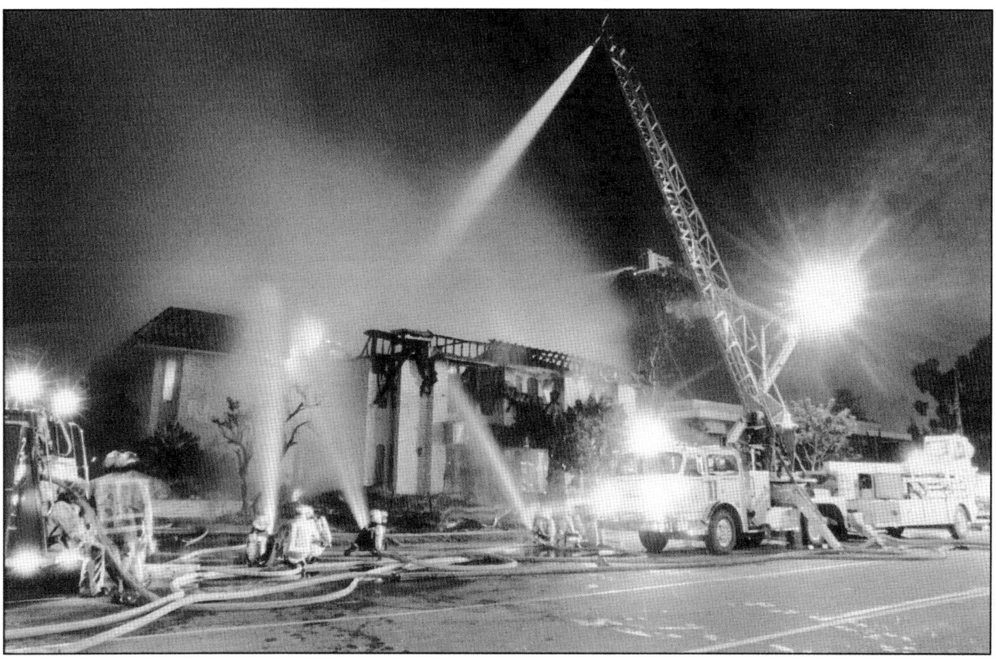

The second National Education Corporation fire occurred in October. Falling debris ruptured a gas line that according to one firefighter, "Went off like a flame thrower," collapsing the roof. This stubborn fire took three hours to extinguish. (Courtesy of Richard Koehler.)

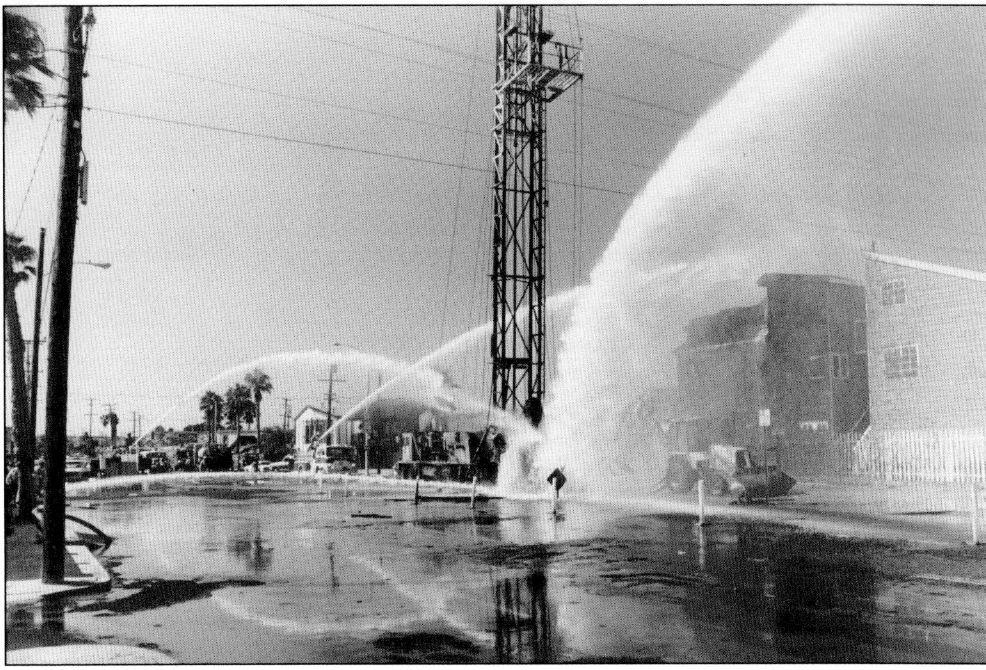

In August 1985, workers attempted to recap a familiar oil well, the same well that caused a fire in 1927 and would prove to be a headache to the NBFD for years afterward. Oil workers heard a large explosion described as a "whoosh." Escaping methane gas ignited and fireballs shot up into the sky. Firefighters kept steady streams of water playing on the well as efforts to stop the surge of oil and gas were underway. (Courtesy of Richard Koehler.)

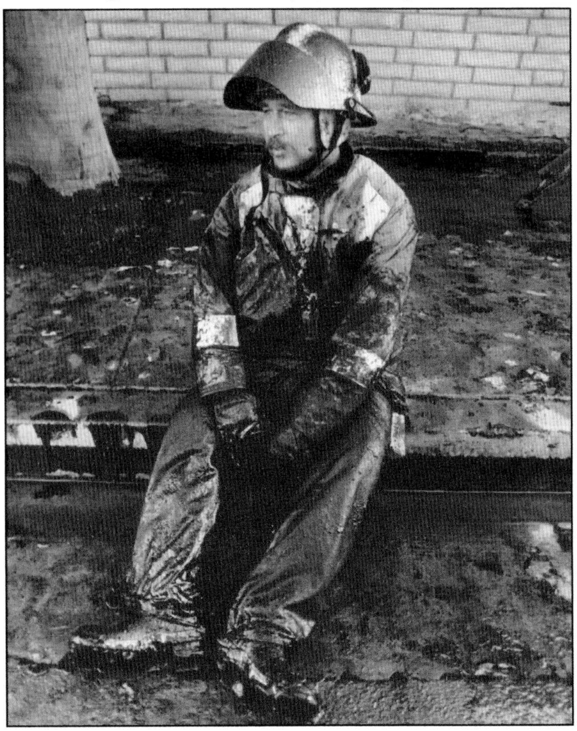

The flow of crude oil and methane gas was eventually stopped five hours after the first explosion by pumping mud into the well. Quick work by firefighters and the U.S. Coast Guard prevented runoff water contaminated with oil from entering the bay. In this photograph, Engineer Doug Wright takes a breather from the relentless shower of crude oil. (Courtesy of Richard Koehler.)

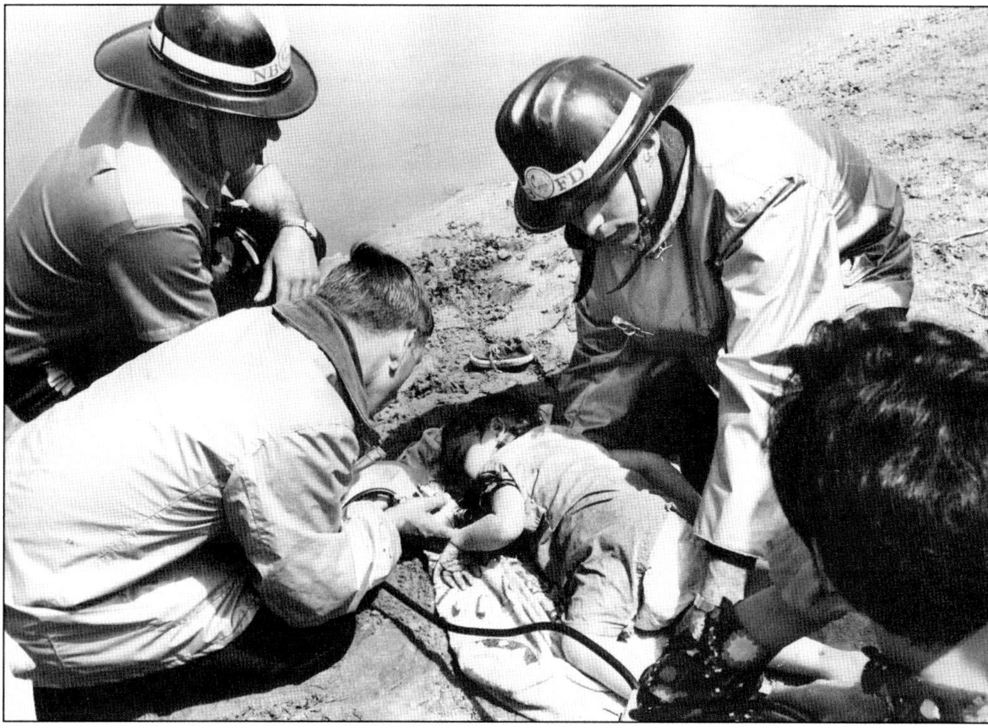

In this photograph, taken around 1970, firefighters administer oxygen to a child on the beach as an anxious parent looks on. Newport Beach lifeguards were reabsorbed into the NBFD in 1995. Today lifeguards are the first responders on scene along the water's edge, responsible for patrolling miles of sandy beaches, rocky shoreline, and harbor waters. (Courtesy of Richard Koehler.)

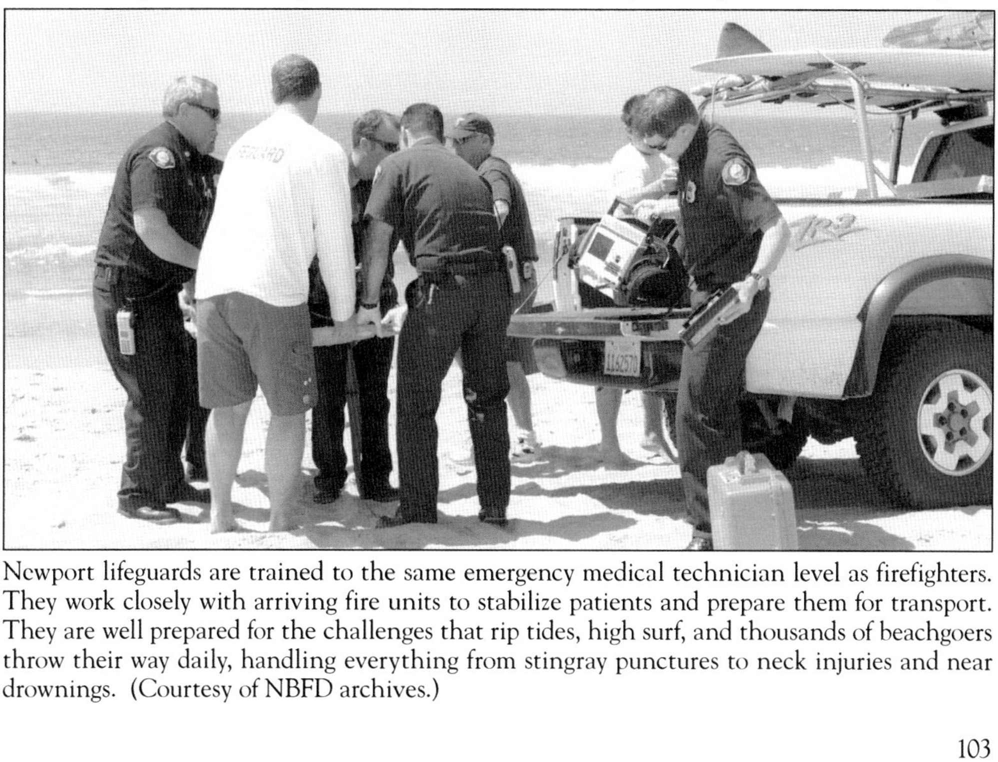

Newport lifeguards are trained to the same emergency medical technician level as firefighters. They work closely with arriving fire units to stabilize patients and prepare them for transport. They are well prepared for the challenges that rip tides, high surf, and thousands of beachgoers throw their way daily, handling everything from stingray punctures to neck injuries and near drownings. (Courtesy of NBFD archives.)

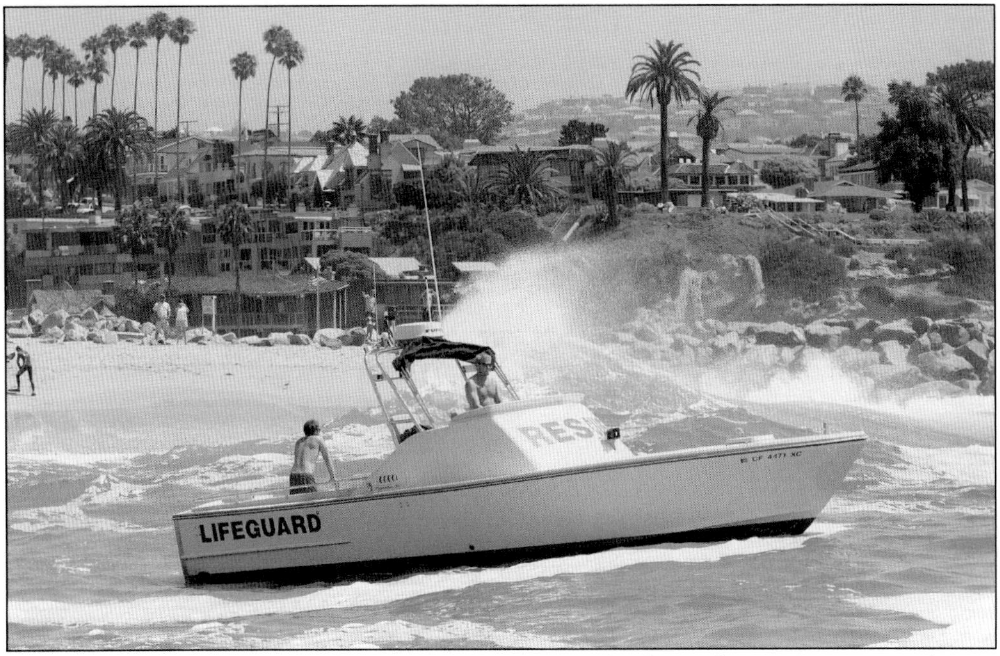

Lifeguard boat operators are adept at maneuvering inside the surf zone if rescue conditions require it. They also have their own dive team. Lifeguards provide sophisticated daily weather and ocean condition updates, continuing the tradition that Chief Frank Crocker started in the 1930s. (Courtesy of NBFD archives.)

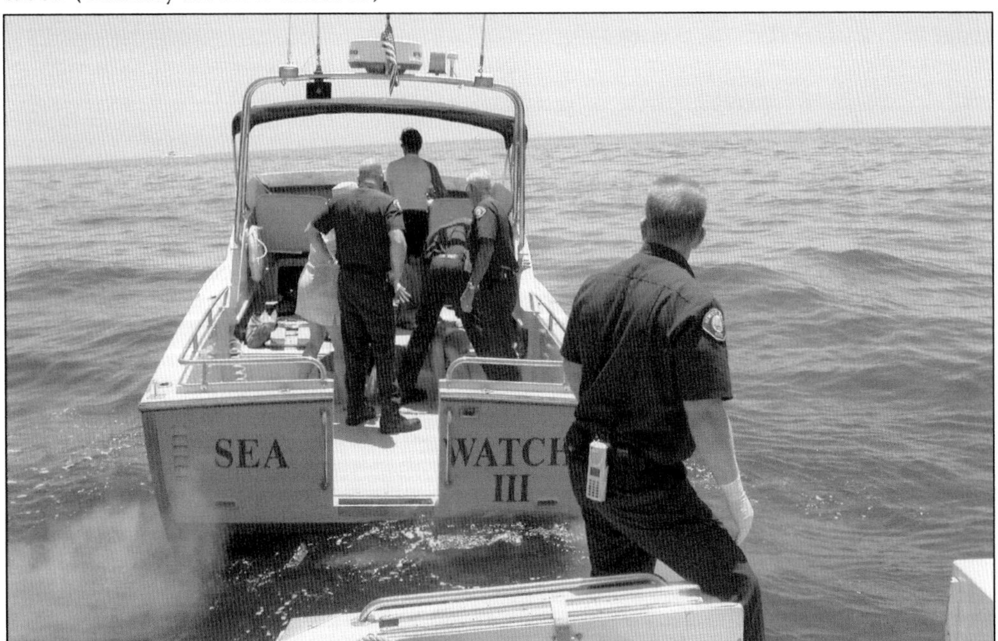

Newport firefighters operate offshore in partnership with lifeguards and Orange County sheriffs. Accidents often occur on boats and critical care must be administered as quickly as possible. Taking emergency care providers directly to the patient, even if it means braving large ocean swells and adverse weather, often means the difference between life and death. (Author's collection.)

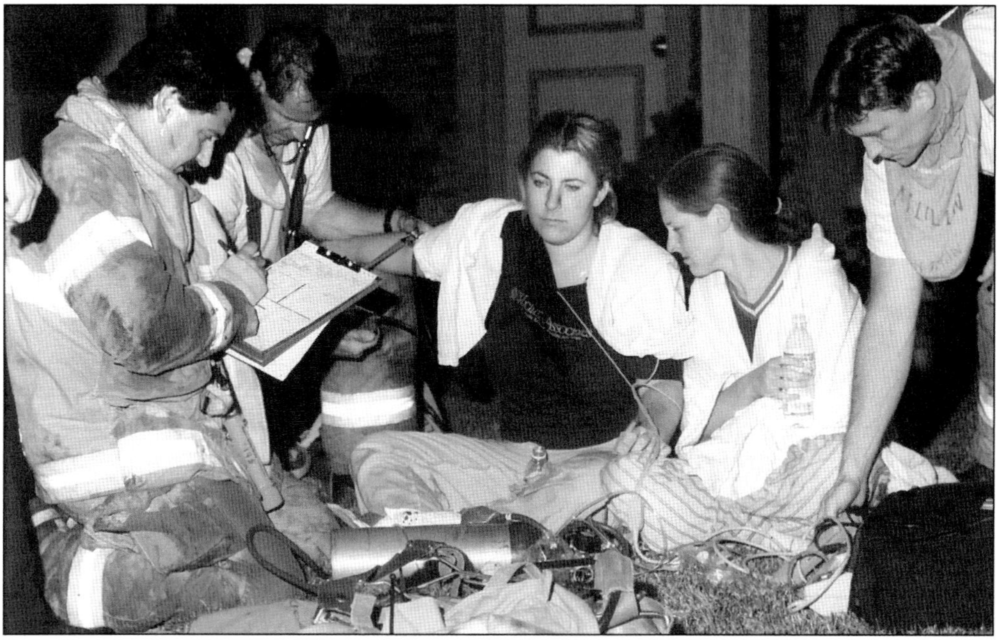

With fire raging below, these two young women found themselves trapped in the upstairs bedroom of their house in May 2000. Smoke forced them to retreat to a bathroom, where they used wet blankets to seal cracks around the door. They stayed in communication with the fire dispatcher until fire burned through their phone line. Firefighters Mike Mullen and Brett Sutherland were able to locate them in the thick smoke and carry them out of the house to safety. (Courtesy of Richard Koehler.)

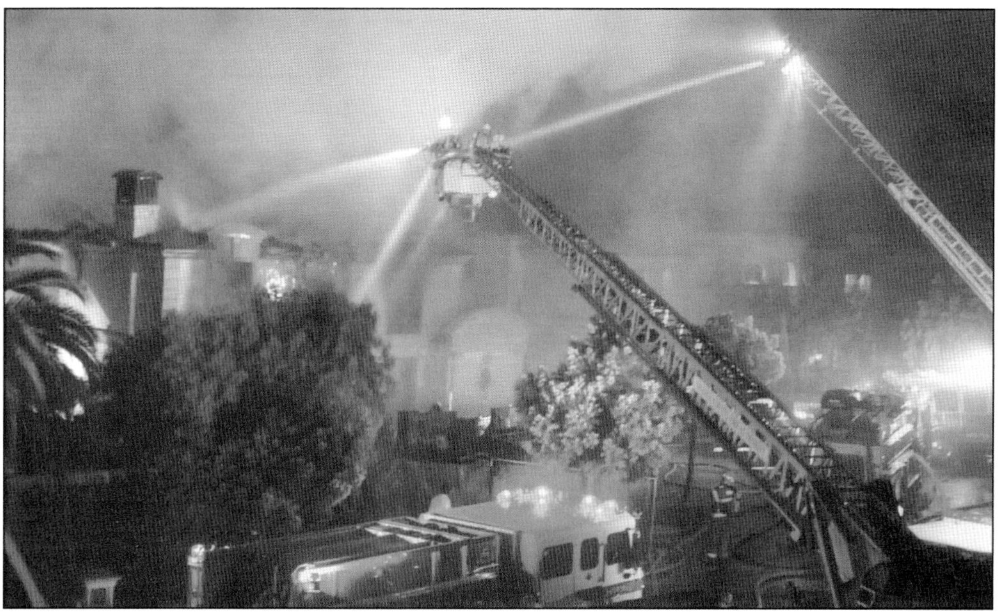

This house fire, in the Newport Coast hills, could be seen from every corner of the city. Two truck companies pour massive streams of water into the house. The truck in the center is an Orange County Fire Authority rig assisting the NBFD under a mutual-aid agreement. (Courtesy of Richard Koehler.)

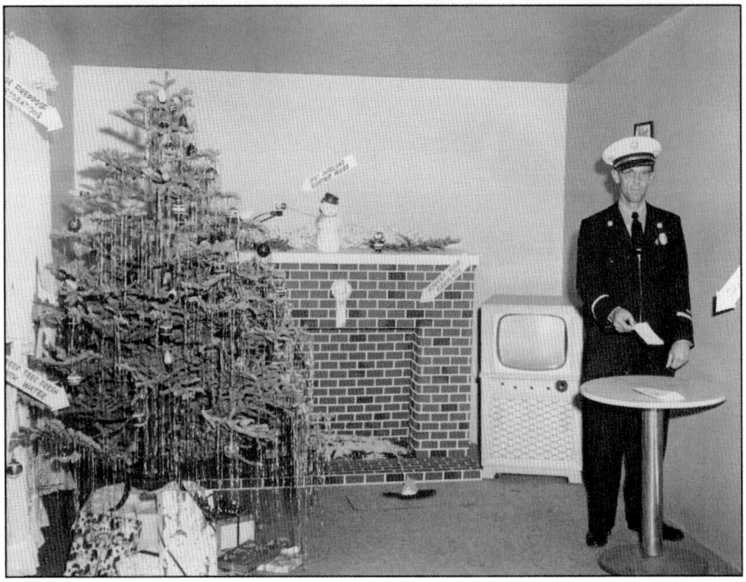

Most people are familiar with annual holiday fire-safety demonstrations put on by local fire departments. In the photograph above, Chief Briscoe offers timely safety tips in December 1952. Ironically, this fire prevention booth was set up in the Rendezvous Ballroom. On Christmas Eve, 54 years later, the house below was destroyed when the family Christmas tree went up in flames. With retirement looming at the end of the year, this was the author's last fire. (Above courtesy of NBFD archives; below courtesy of Richard Koehler.)

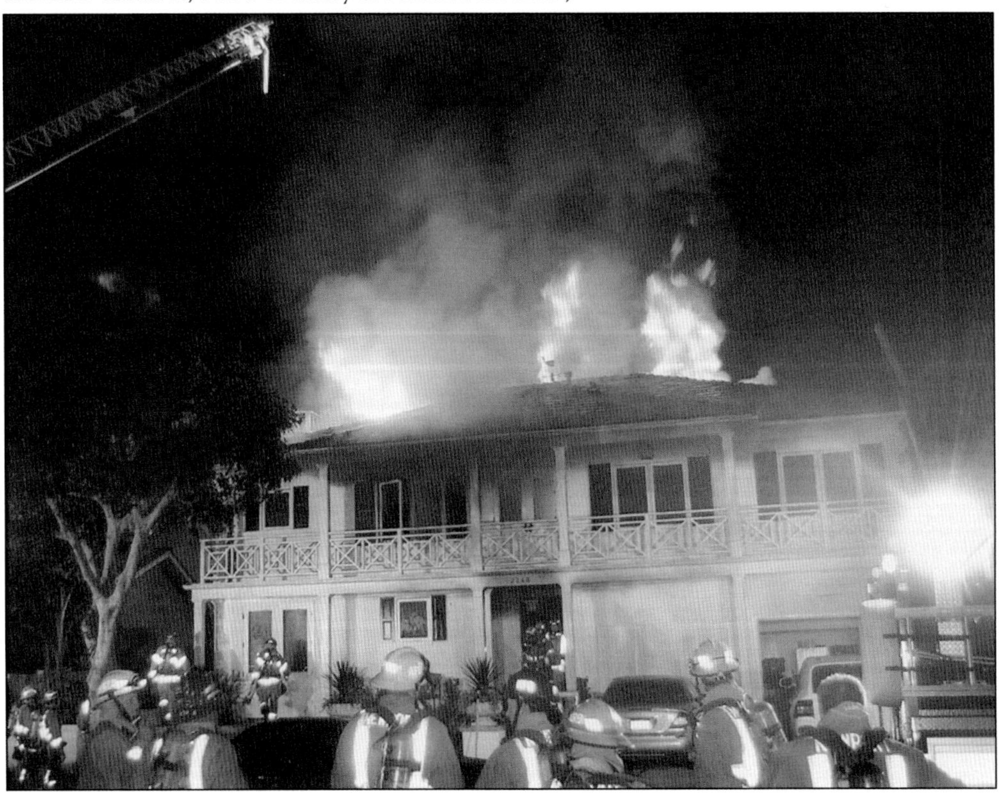

Five

FIGHTING FIRE

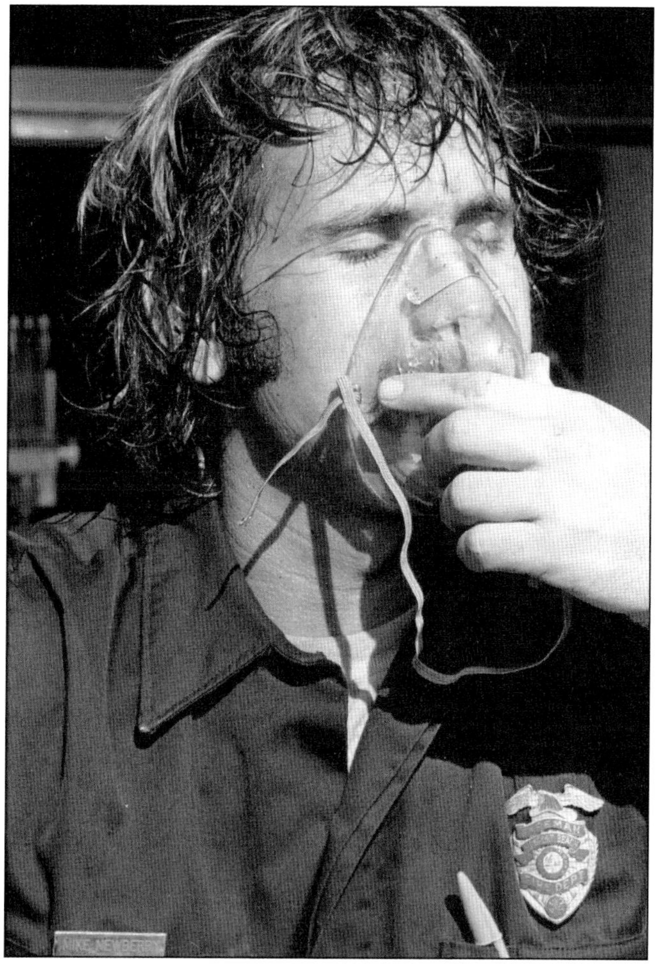

Firefighter/paramedic Mike Newberry takes a well-earned breather while fighting a fire on a hot summer day. Throughout its history, the NBFD has been committed to public fire-prevention education and enforcement of the latest fire and building codes. The Fire Prevention Bureau has worked tirelessly to make the residents of Newport Beach as safe as possible. Fires do occur from time to time, however. Fighting and extinguishing fire is exhausting and dangerous work. When called, the men and women of the NBFD go about this pursuit with a dedication and purpose second to none. (Courtesy of NBFD archives.)

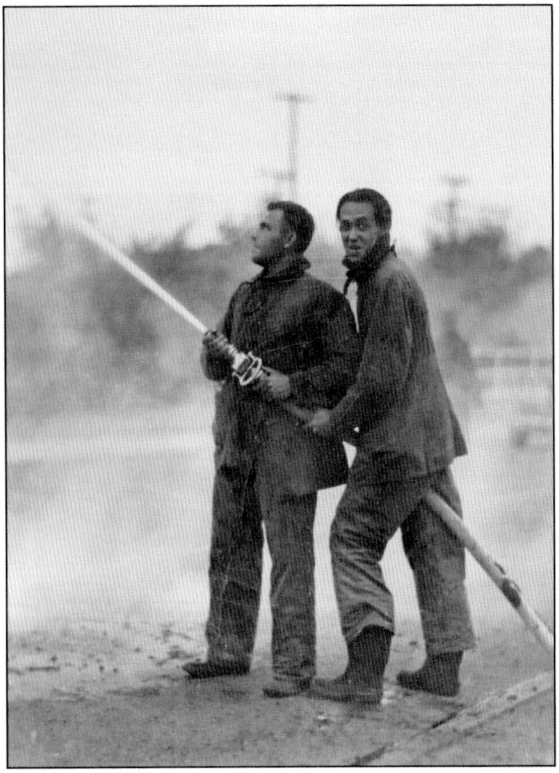

Volunteer firefighters Barney Leyman and Harold Caldwell wrestle a hose line outside the Rendezvous Ballroom in 1935. They wore primitive canvas turnouts and no helmets. They probably received only cursory training. Their enthusiasm and dedication carried them through the day. (Courtesy of NBFD archives.)

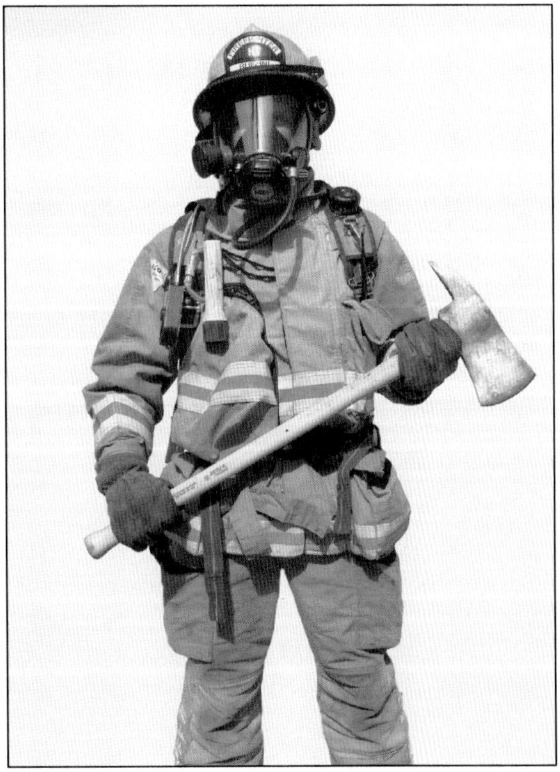

Today's Newport Beach firefighter is clad head to toe in the latest protective clothing. Each firefighter carries a radio, pager, self-contained breathing apparatus with a personal alarm locator device and face mask, and an assortment of tools and equipment. No firefighter enters a hazardous environment without this gear in place and operational. (Author's collection.)

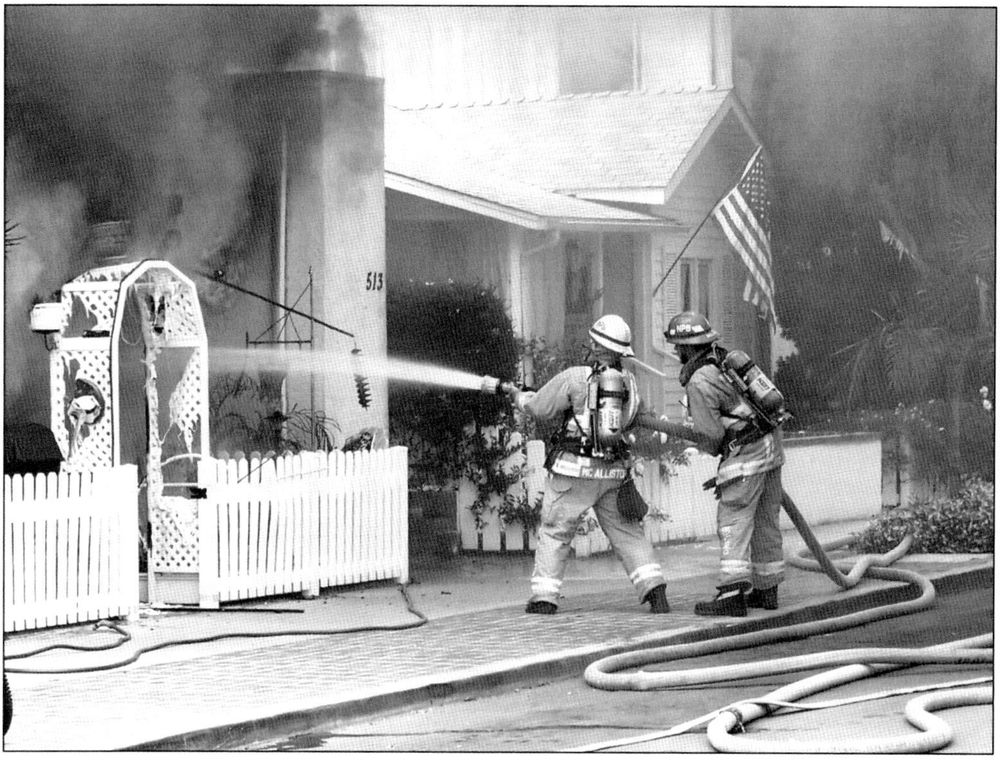

Rescue is always preeminent in the minds of firefighters. When occupants are known to be out of the house and safe, attacking and extinguishing the fire become priorities. In this photograph, firefighters attempt to achieve a quick knockdown of a house fire from outside. (Courtesy of Brent Jacobsen.)

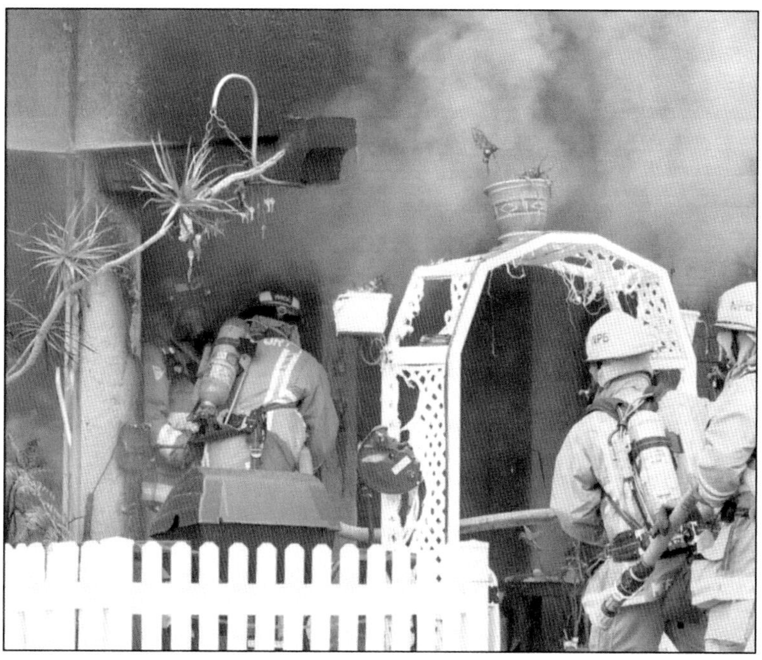

Newport Beach firefighters move in to make an interior attack. Firefighters enter as a team. No one goes in unless the hose line is charged, all safety gear is donned correctly, and breathing apparatus is on. Outside, a safety team is in place with a second hose line should its assistance be needed. (Courtesy of Brent Jacobsen.)

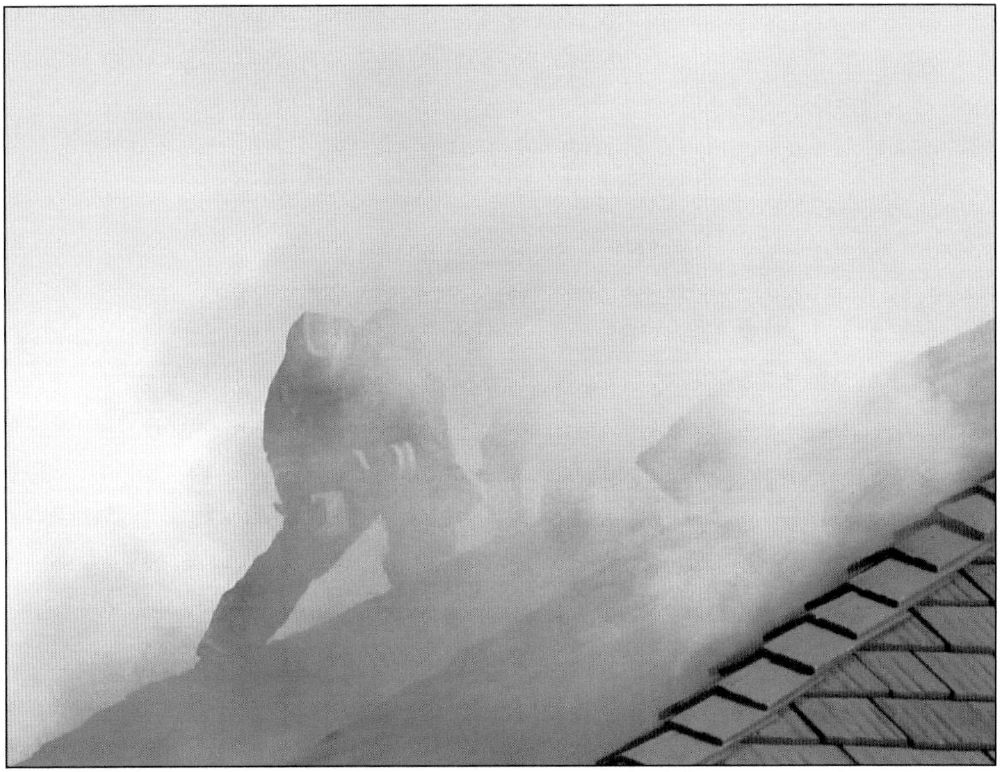

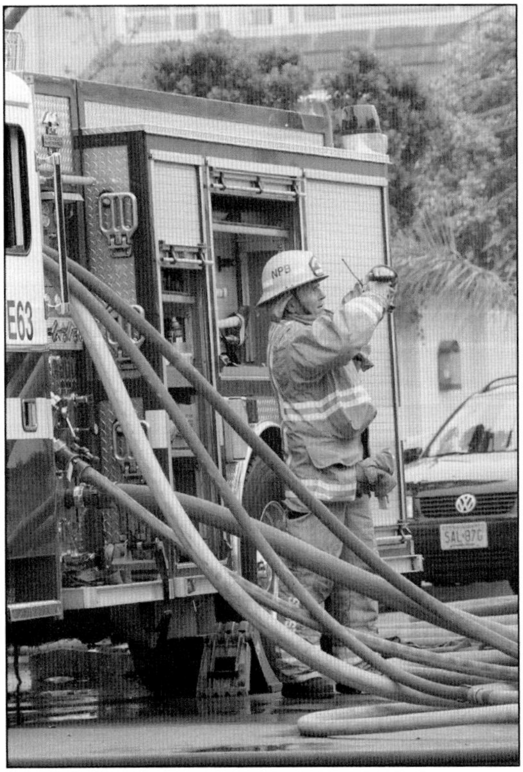

NBFD firefighters cut ventilation holes in a roof as the fire rages below. These holes allow heat and smoke to escape the interior, making it safer for firefighters to enter and extinguish the fire. (Courtesy of Jen Schulz.)

Water extinguishes fire. It is the engineer's responsibility to provide water to firefighters on the nozzle. At large fires, securing a water supply, keeping track of hose lines, and maintaining proper pressures to insure firefighter safety and effectiveness make the engineer's position one of the most important responsibilities on the fire ground. (Courtesy of Brent Jacobsen.)

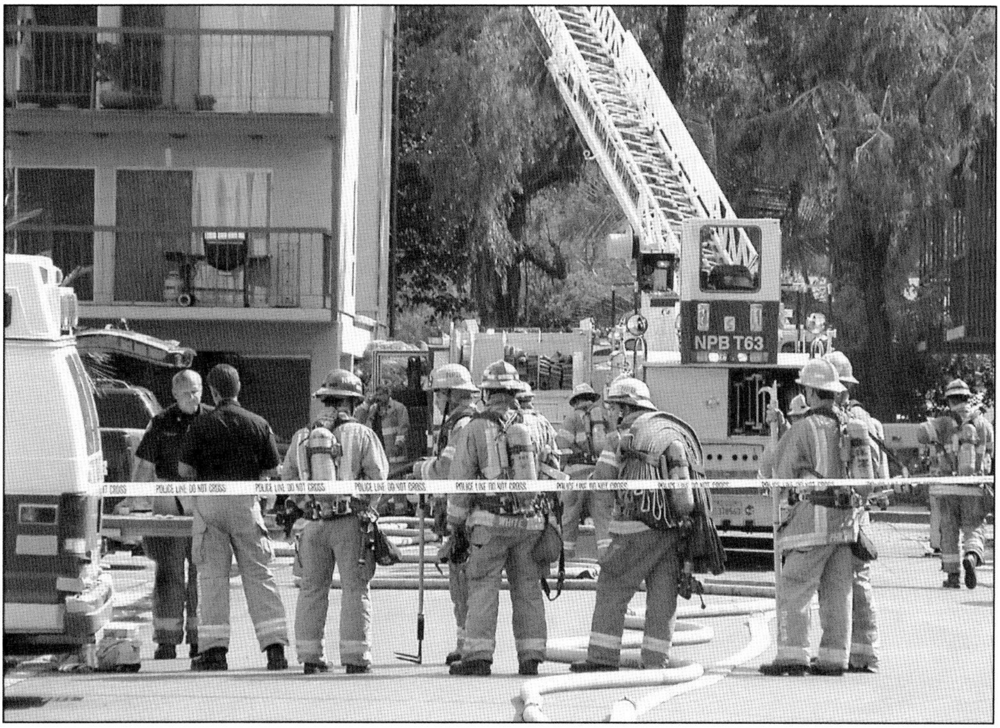

The NBFD believes that the number-one priority on the fire ground is firefighter safety. Adequate personnel are always on hand in staging areas to provide enough manpower to handle any problem that may arise, replace exhausted crews, and to be in position to assist crews that may find themselves in trouble. (Author's collection.)

Catastrophic wildfires periodically sweep through Southern California. Multiple strike teams, each composed of five fire engines and a battalion chief, are dispatched to these fires whenever homes are threatened. Their job is to protect these homes and other structures endangered by the blaze. (Author's collection.)

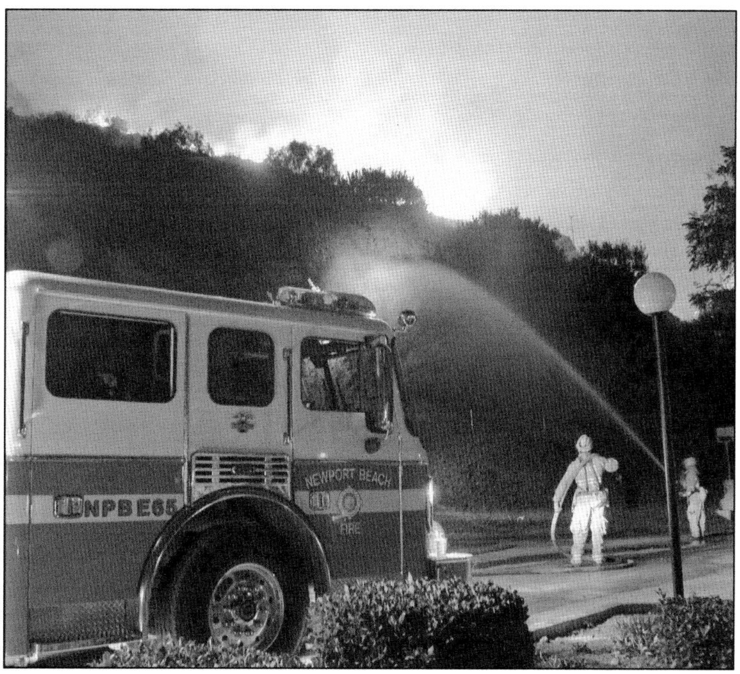

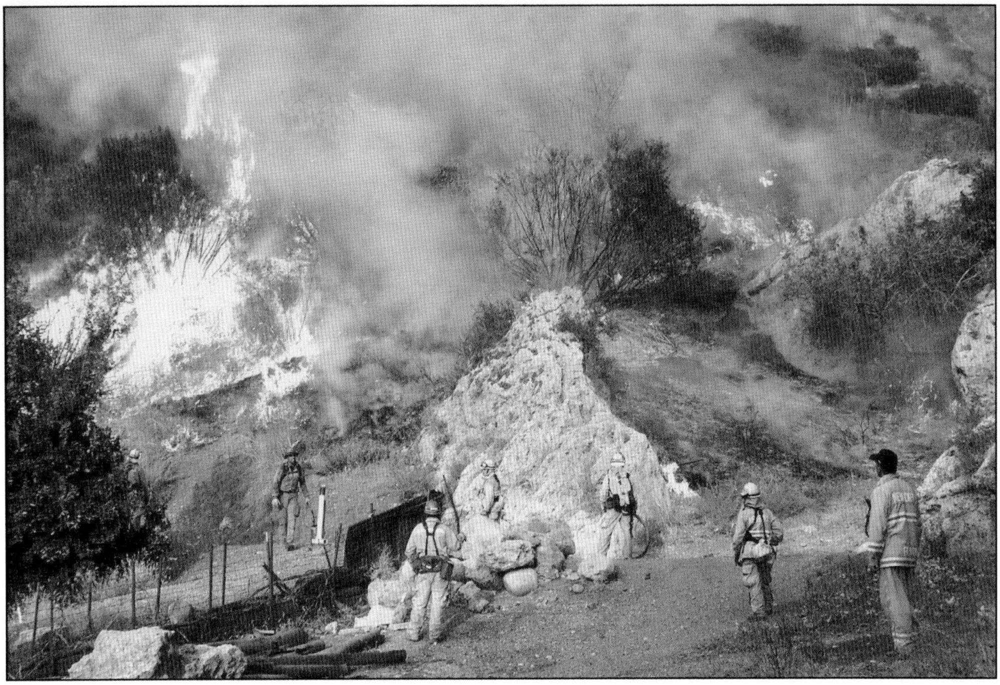

Strike team crews often stay out for up to five days and range from San Diego to Santa Barbara. Wildfires that encroach on urban areas present firefighters with new and unique firefighting challenges. Newport Beach Fire sent engines to the Malibu and Laguna Beach wildfires in 1993 and to all the large fires that have occurred since. (Author's collection.)

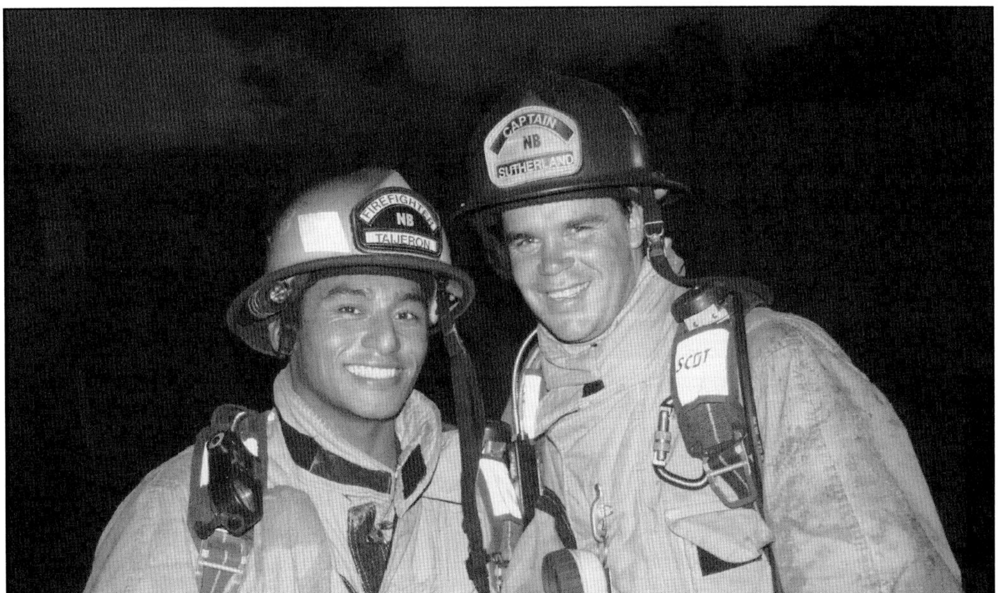

Roman Taijeron and Brett Sutherland put a face on the feeling of personal satisfaction after a job well done. Preventing the spread of fire or stabilizing a life-threatening injury or illness makes the profession of firefighter one of the most rewarding and exciting careers a man or woman can pursue. (Author's collection.)

Six

100 Years of Community Service

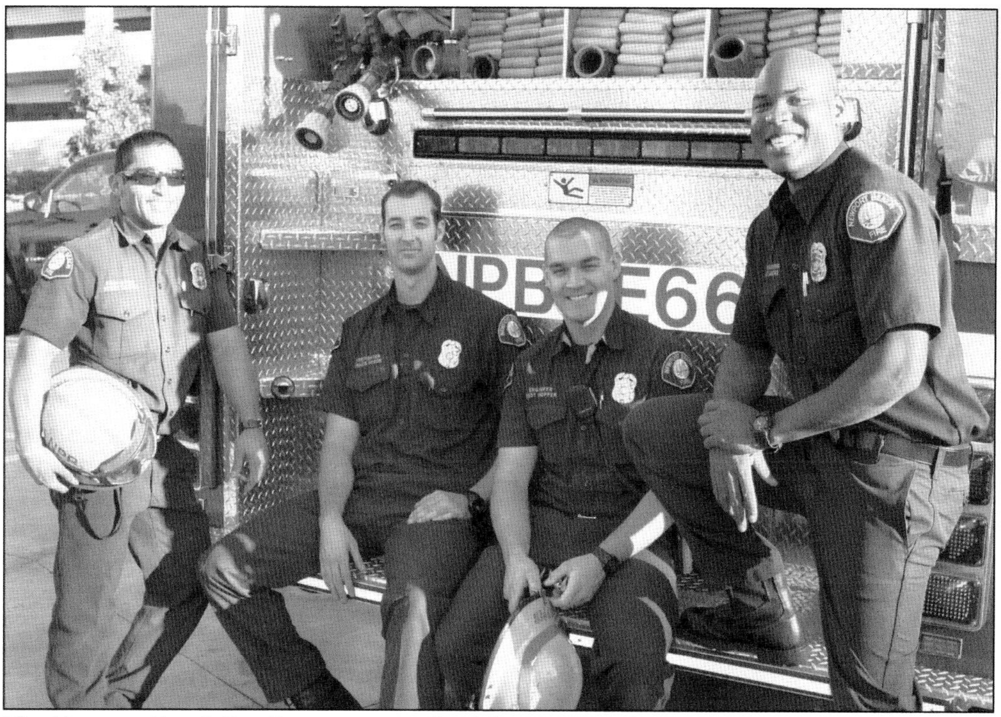

The Newport Beach Fire Department will celebrate its 100th anniversary in April 2011. Poised on the brink of this milestone, the department has never been more capable of serving the ever-growing city of Newport Beach, nor better prepared for the contingencies modern times may throw its way. In addition to the men and women who man the eight fire stations and the lifeguard boats and lifeguard towers, a dedicated group of professionals staff a fire-prevention bureau, create and implement disaster preparedness and Community Emergency Response Team programs, administrate the paramedic transport program, and insure that the training of all personnel is maintained at an exemplary level. (Author's collection.)

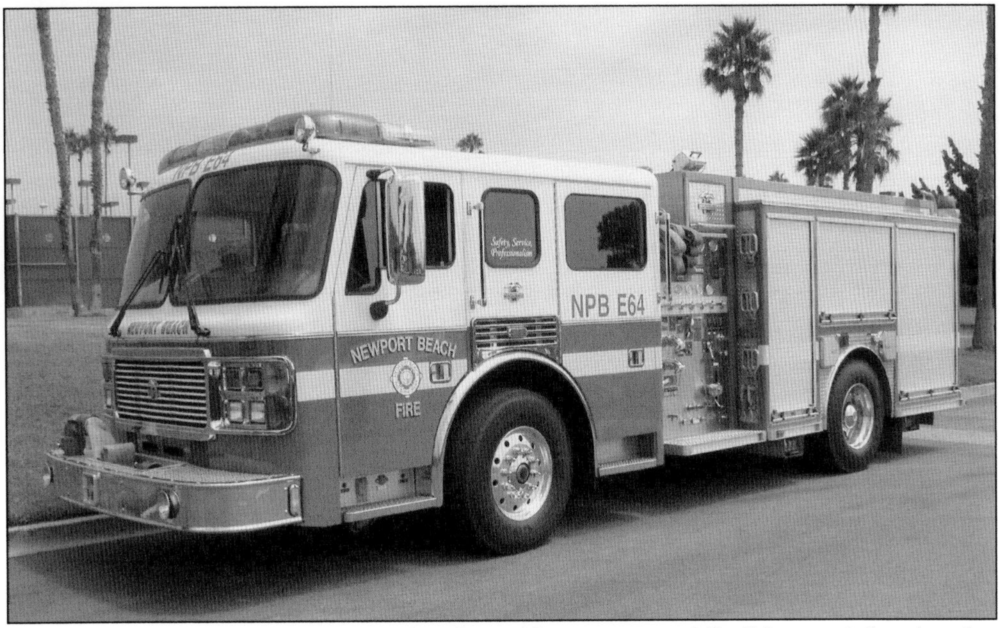

The NBFD maintains a modern, first-class fleet of fire apparatus. All of Newport's frontline fire engines are American LaFrance Eagles identical to this one. In addition to carrying 500 gallons of water and hose and first-aid equipment, these engines are equipped for extended campaigns throughout Southern California on brush-fire strike teams. (Author's collection.)

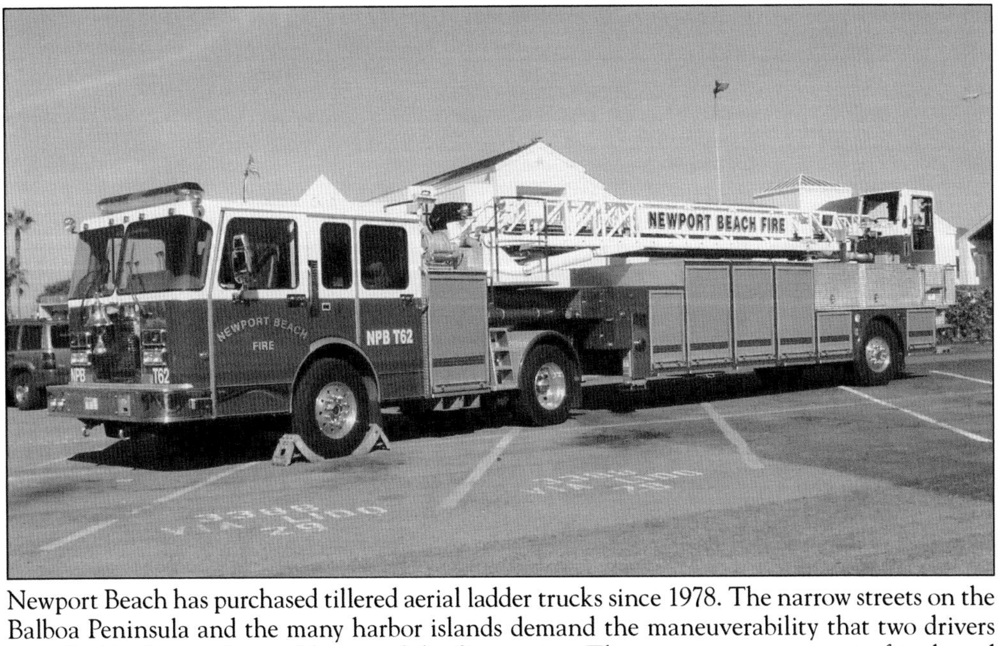

Newport Beach has purchased tillered aerial ladder trucks since 1978. The narrow streets on the Balboa Peninsula and the many harbor islands demand the maneuverability that two drivers provide. Trucks are the workhorses of the fire service. They carry an assortment of tools and equipment necessary for interior and exterior work at fire scenes and auto accident extrication. (Author's collection.)

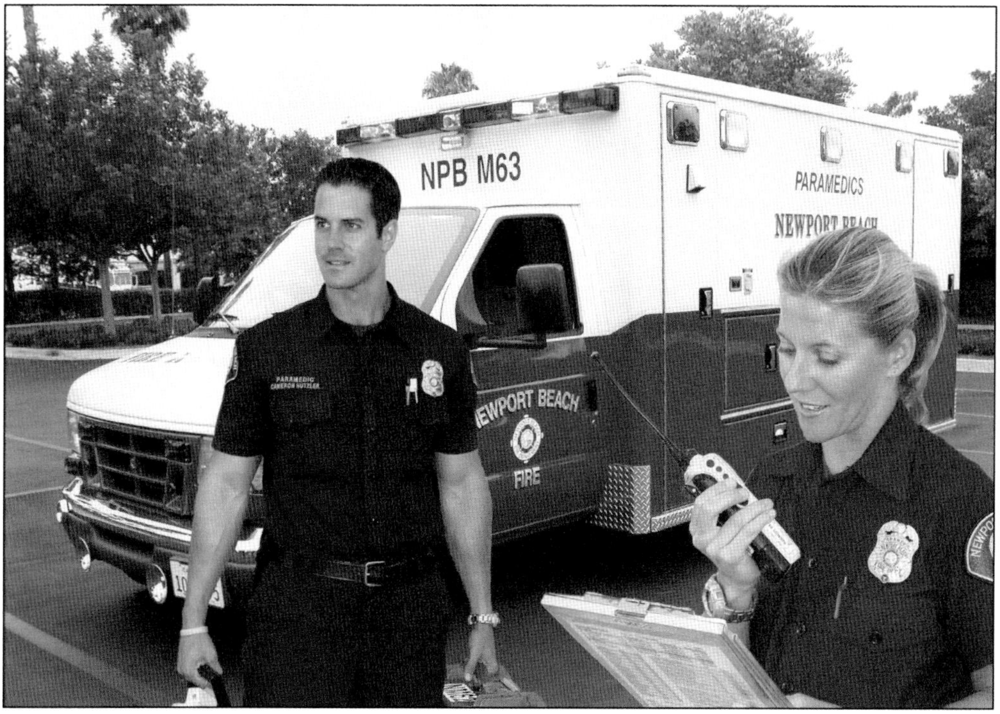

Since 1996, paramedics, like Cameron Hutzler and Kelly Gorman seen here, have been transporting patients directly to the hospital. This innovative program has insured that the Newport Beach Fire Department will provide continuous professional care and treatment from the moment of patient contact to the hospital emergency room. (Author's collection.)

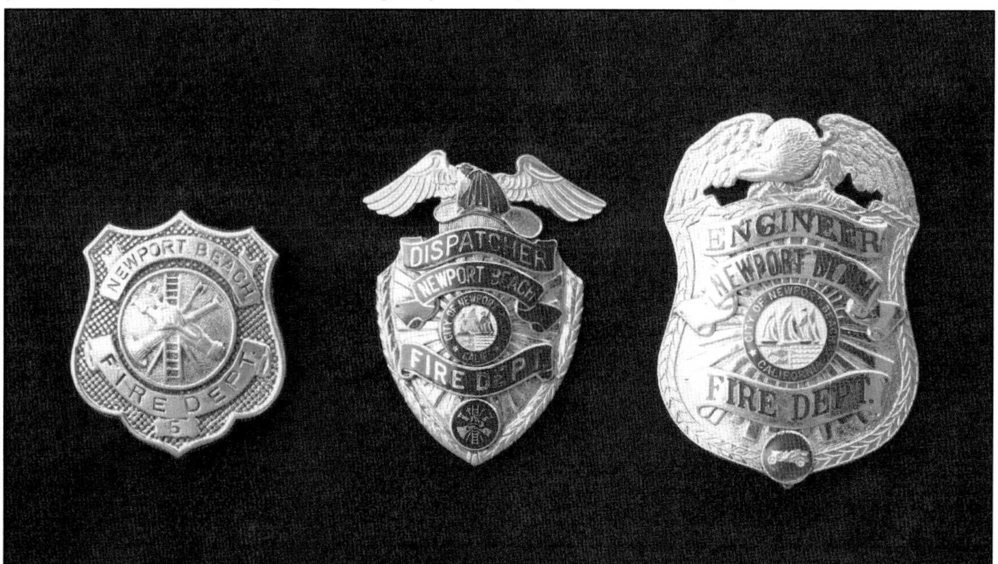

The badge worn by Newport Beach firefighters has changed dramatically in 80 years. First introduced in 1927, the badge denotes affiliation, rank, and authority. In Newport Beach, it also testifies to the oath sworn by each man and woman to serve the community above and beyond the call of duty. (Author's collection.)

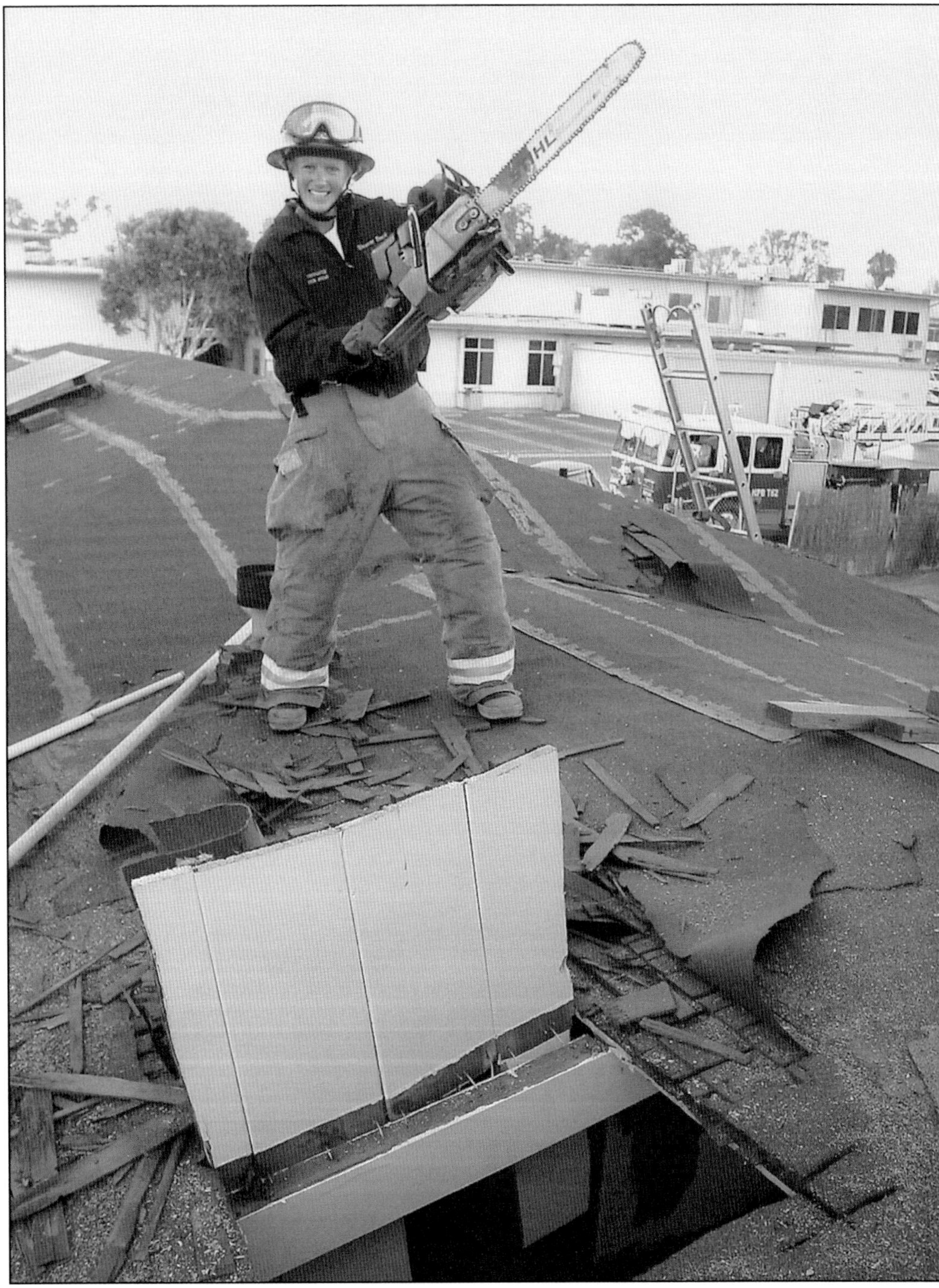

Paramedic Erin Brown cuts a perfectly louvered ventilation hole in the roof of an abandoned building. Training never stops. It remains an ongoing process, whether it involves practical exercises like this one, classroom instruction, or reading the latest professional trade magazine. A new fire station has been built near John Wayne Airport with training in mind. In addition to classroom facilities, a four-story training tower has been constructed on the site. (Author's collection.)

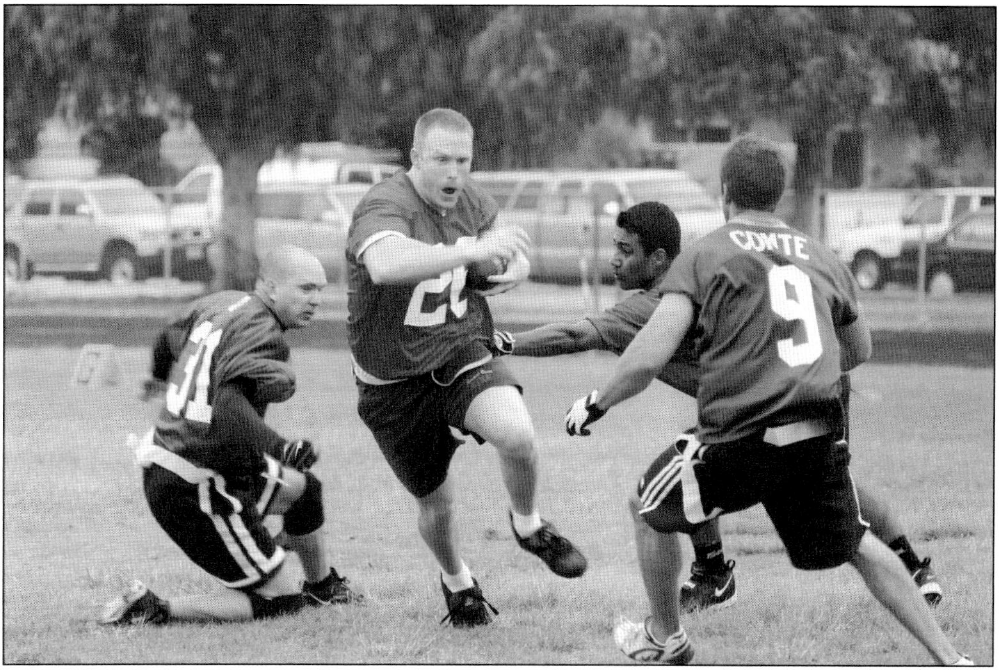

A mutual respect and a healthy rivalry have always existed between Newport Beach fire and police personnel. A younger generation of firefighters and police officers has taken over the helm in the 21st century. They have used their youthful athleticism and fitness as a way of further solidifying the bond between safety personnel. In the photograph above, firefighter Casey Blythe displays Heisman Trophy form. (Both author's collection.)

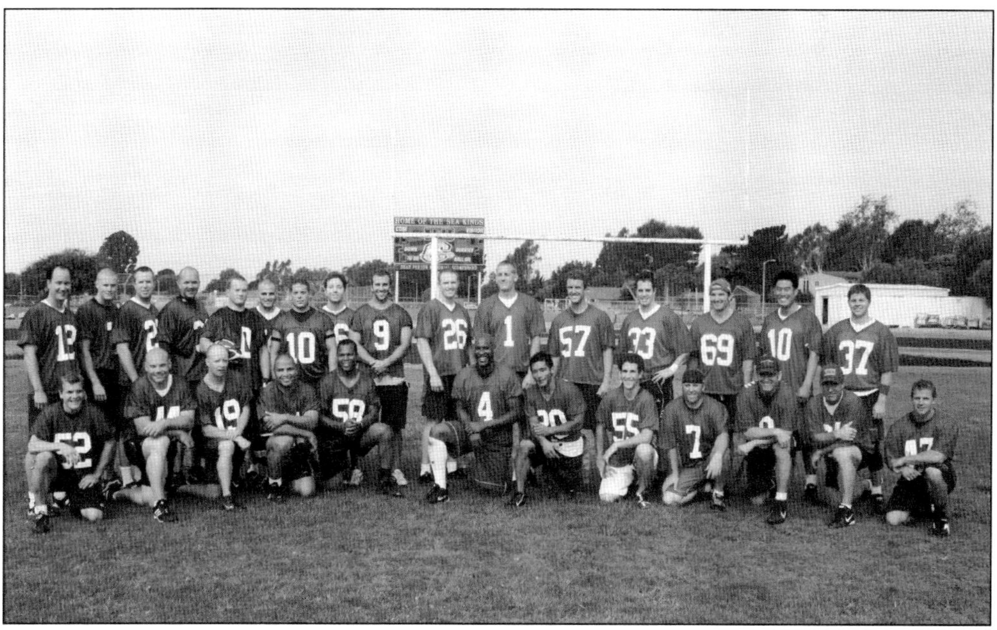

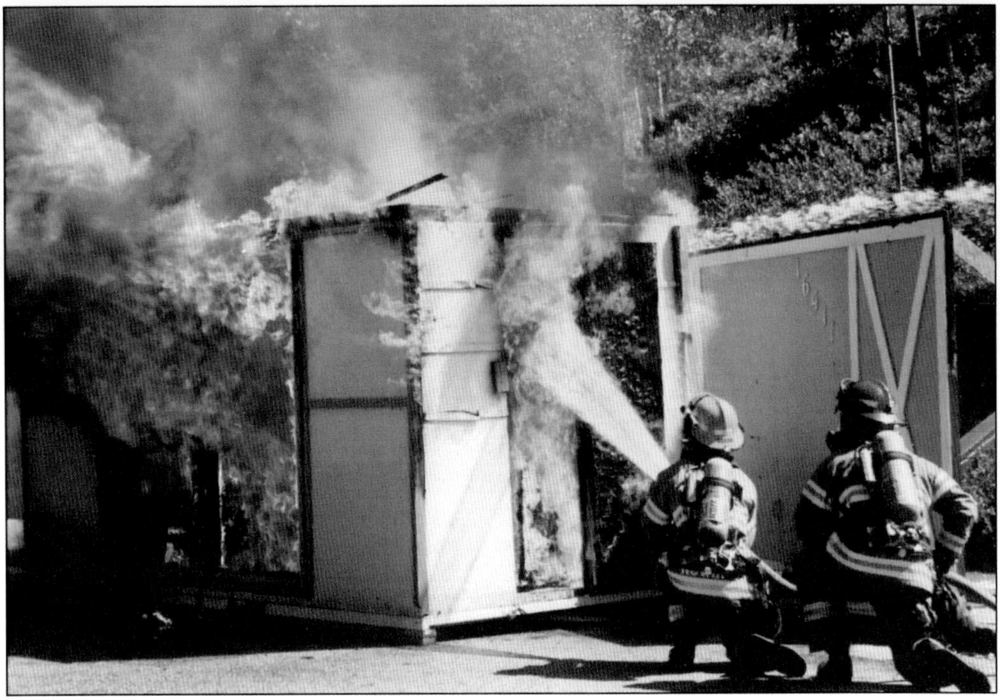

Seventy-five years after Chief Crocker suggested that city residents visit their local fire stations, open house, or Public Safety Day as it is now called, continues as an annual tradition. The fire department, police department, and lifeguards put on a show that increases in popularity with each passing year. (Courtesy of NBFD archives.)

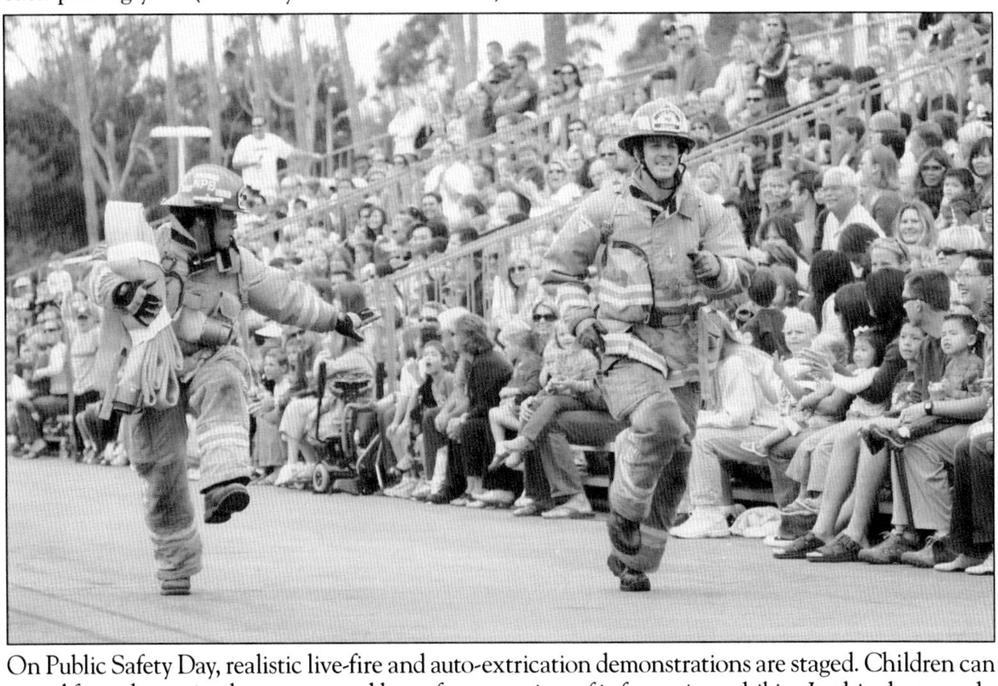

On Public Safety Day, realistic live-fire and auto-extrication demonstrations are staged. Children can rappel from the station hose tower and learn from a variety of informative exhibits. In this photograph, a little friendly competition between firefighters delights the crowd. (Author's collection.)

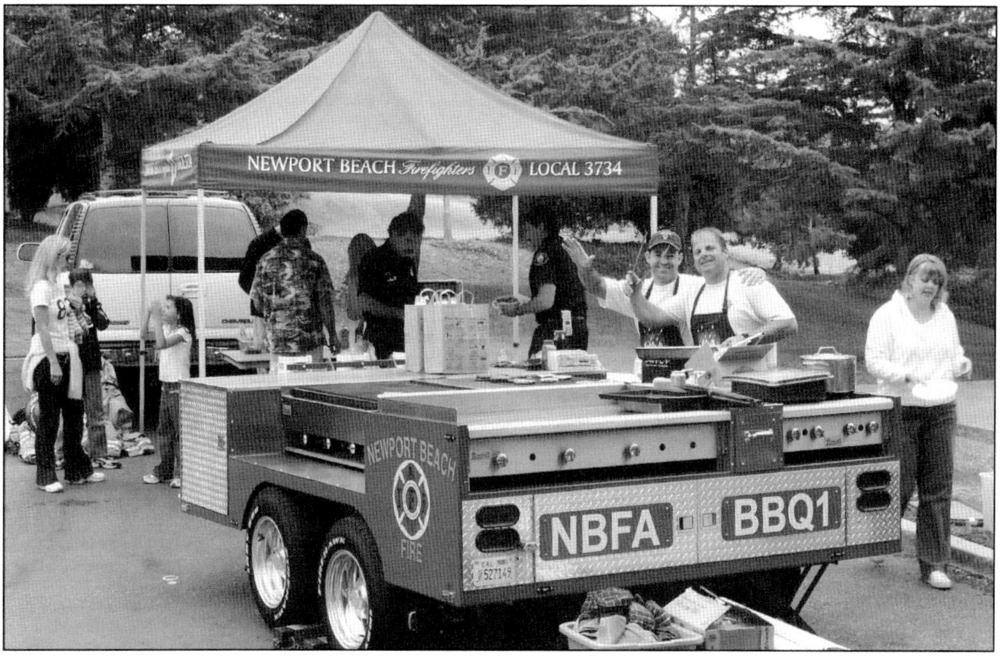

Community leaders teamed up with the Newport Beach Firefighters Association to design, fund, and build a state-of-the-art barbecue on wheels. The association's best chefs, like Carlos Medina and Jim Jeziorski pictured here, give the barbecue a workout each Public Safety Day and at community events throughout the year. (Courtesy of Jeff Remmele.)

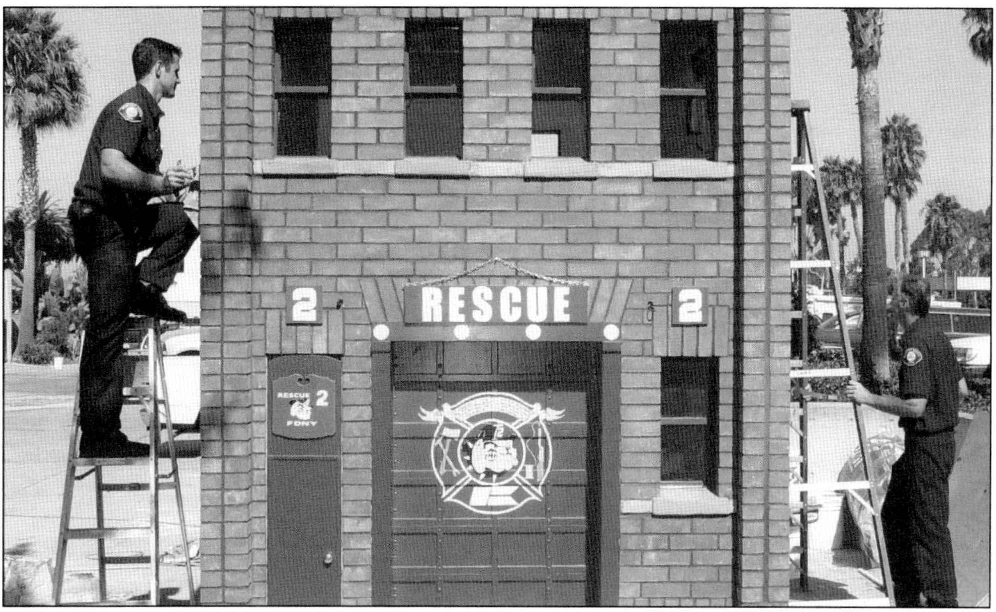

The Newport Beach Firefighters Association has always been active in aiding charities like the Make-A-Wish Foundation and the Orange County Burn Center. Over the years, the association has constructed three playhouses for auction by HomeAid, an Orange County-based charity. The playhouse in this view was a scaled-down copy of the FDNY's Rescue Two fire station in Brooklyn. Rescue Two lost seven men on September 11. (Author's collection.)

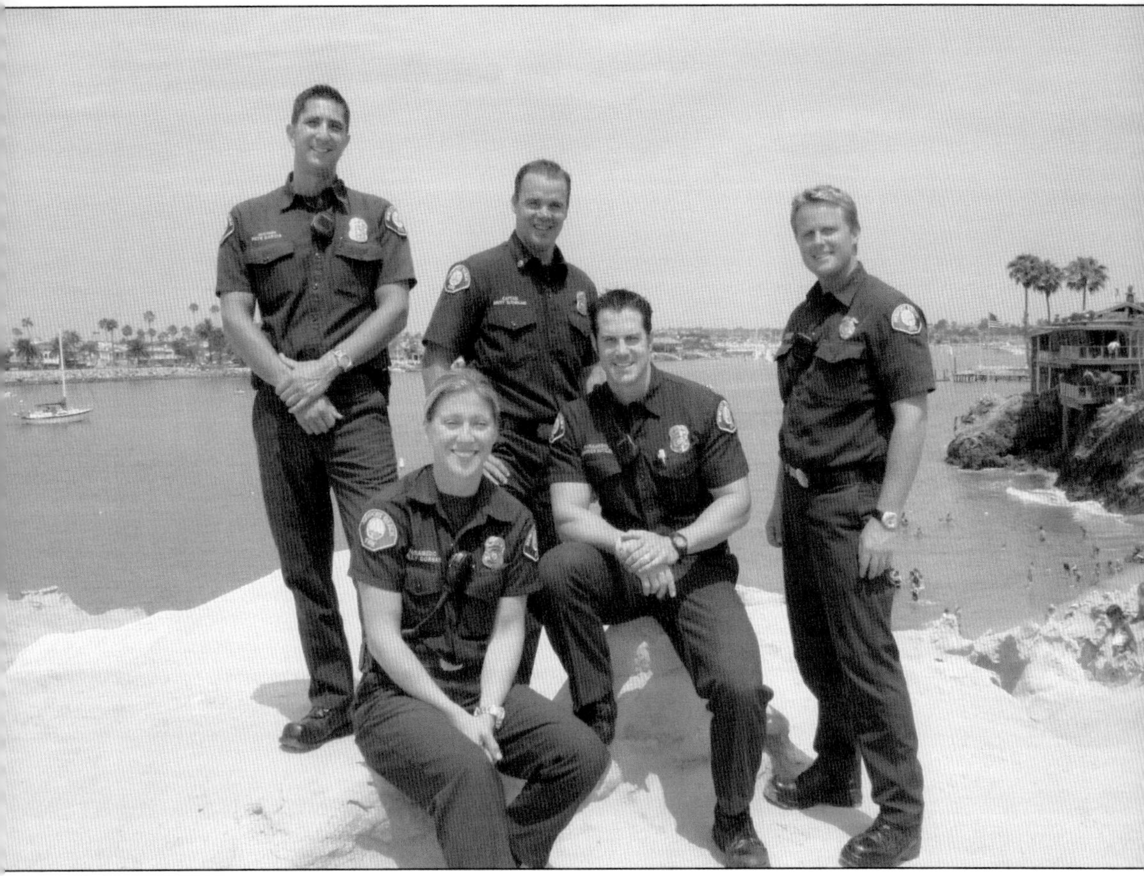

Pictured from left to right, Newport Beach firefighters Pete Garcia, Kelly Gorman, Brett Sutherland, Cameron Hutzler, and Jim Reidler stand above the entrance to Newport Harbor. Ninety-five years before this photograph was taken, Margaret Frost jumped into the water near this very spot to rescue her son Lowell Jr. Both perished. Today the men and women of the NBFD build on the tradition of excellence exemplified by all past generations who have worn the badge. They selflessly strive, through extensive training, diligence, perseverance, and sacrifice, to insure that tragedies like the one that befell the Lowell family will not happen on their watch. (Author's collection.)

Seven

THE FIRE CHIEFS

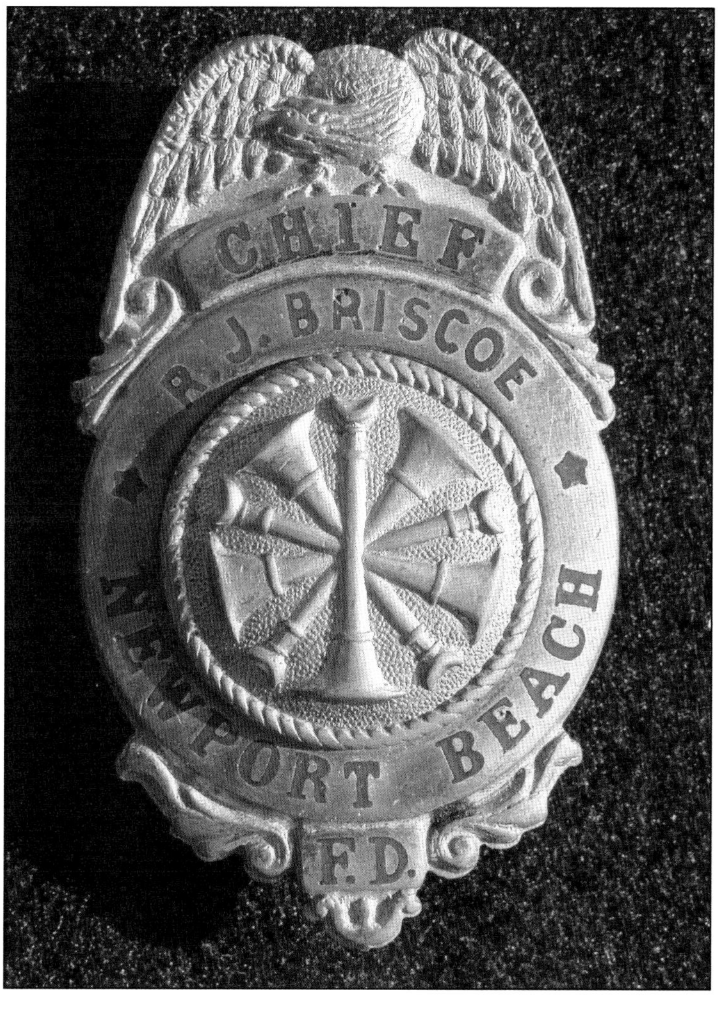

The fire chief, as set forth in Ordinance No. 315, is charged with the discipline and efficiency of the fire department of the City of Newport Beach and is responsible for the regulation, government, and control of the entire department. The Newport Beach Fire Department has been fortunate; only six men have served as fire chief, providing stabilization, a shared vision for the future, a commitment to continued excellence, and a wellspring of organizational pride that flows through the ranks. (Courtesy of NBFD archives.)

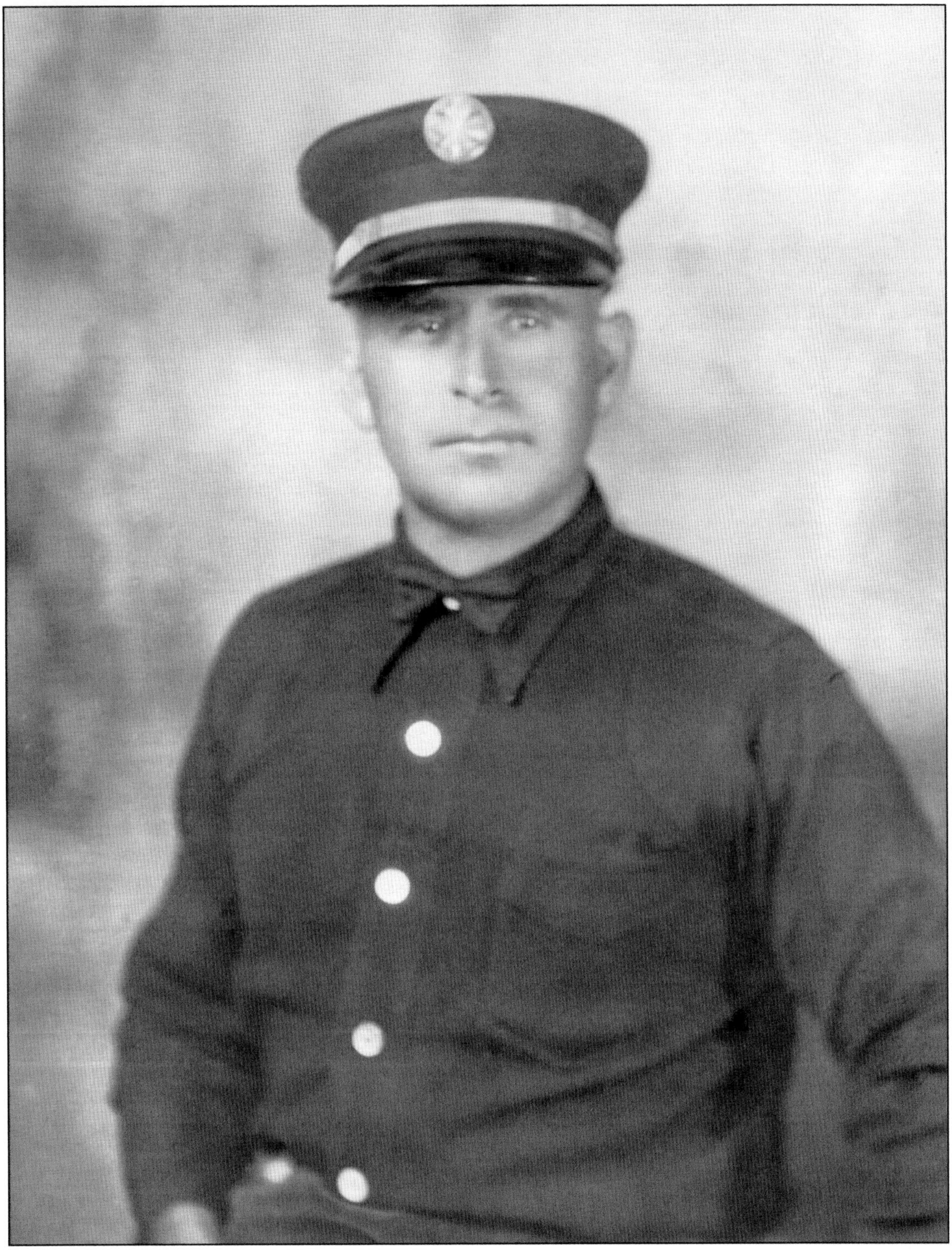

FRANK CROCKER: 1927 TO 1952. Frank Crocker, an electrician by trade, was appointed as the first paid fire chief of the Newport Beach Fire Department in 1927. He immediately set about undertaking a complete reorganization and modernization of the department. During Crocker's 25-year career, three new fire stations were constructed. Each station contained a new fire engine and was manned with at least two paid firefighters. Under Chief Crocker's leadership, the lifeguard corps was professionalized. Public fire-prevention awareness became a priority. Chief Crocker encouraged the public to interact with their neighborhood fire station and firefighters by establishing an annual open house day. (Courtesy of NBFD archives.)

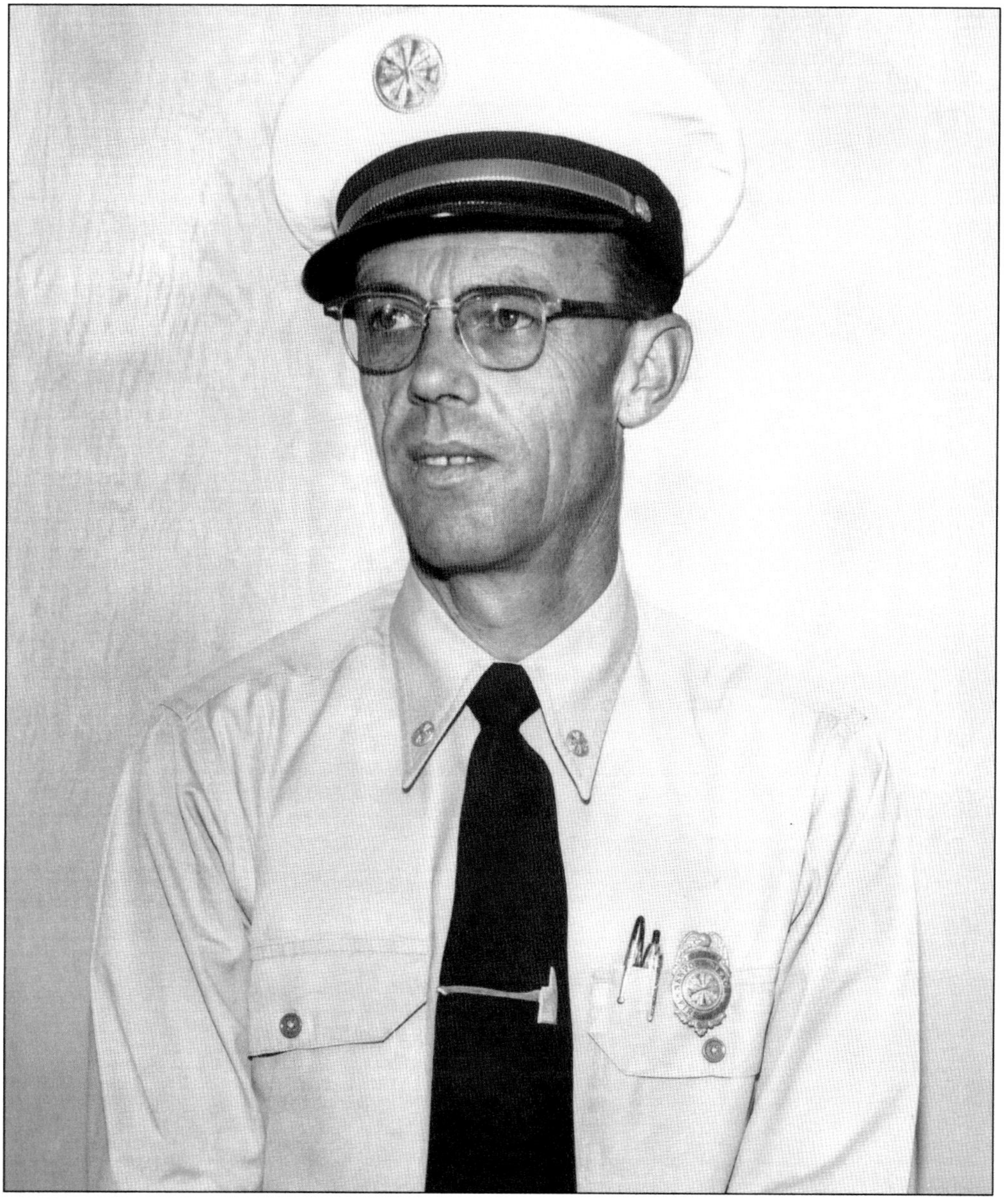

JAN BRISCOE: 1952 TO 1972. Rufus Janvier Briscoe worked as an apprentice electrician under Frank Crocker. When Chief Crocker was appointed fire chief in 1927, Jan Briscoe signed on as a volunteer. Four years later, he was the third firefighter added to the payroll. When Chief Crocker retired in 1952, assistant chief Briscoe was appointed fire chief. During Chief Briscoe's term, the department was expanded to six fire stations, including a new headquarters station. Two truck companies were added, and manpower was increased to implement the formation of three-man engine companies. Dispatching was improved, modern fire apparatus was purchased, and public fire-prevention awareness was heightened. (Courtesy of NBFD archives.)

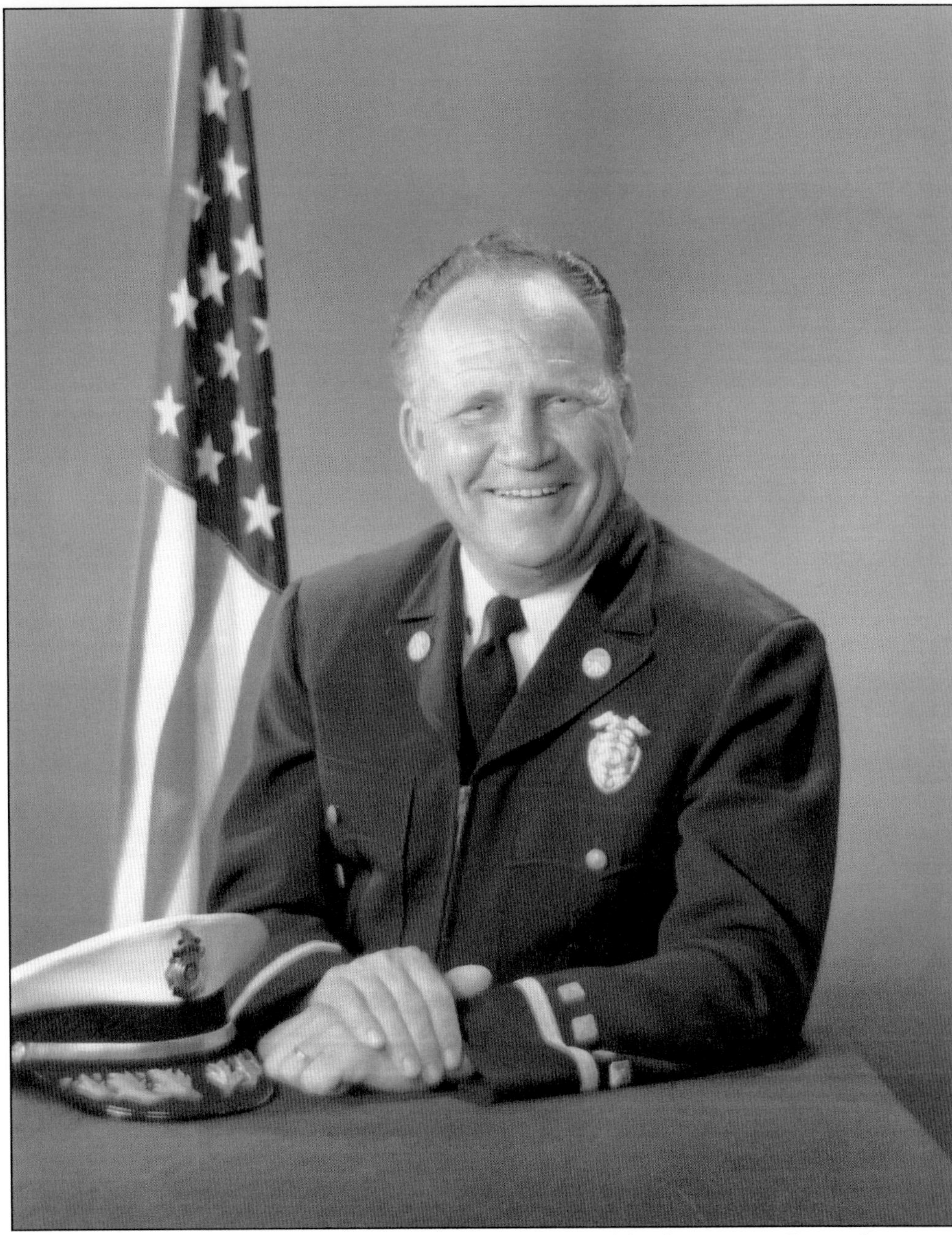

LEO LOVE: 1972 TO 1978. Leo Love was a 28-year veteran of the department. During his tenure as fire chief, the paramedic program was inaugurated and the 911 emergency telephone system was put in service. Chief Love, along with other Orange County fire chiefs, developed the Orange County Mutual Aid Plan, assuring that any city in need of additional firefighting resources would receive them immediately from neighboring fire departments. Training standards were established and building codes were strengthened. Chief Love also purchased a Jaws of Life, a hydraulic tool essential for extrication of victims in automobile accidents. (Courtesy of NBFD archives.)

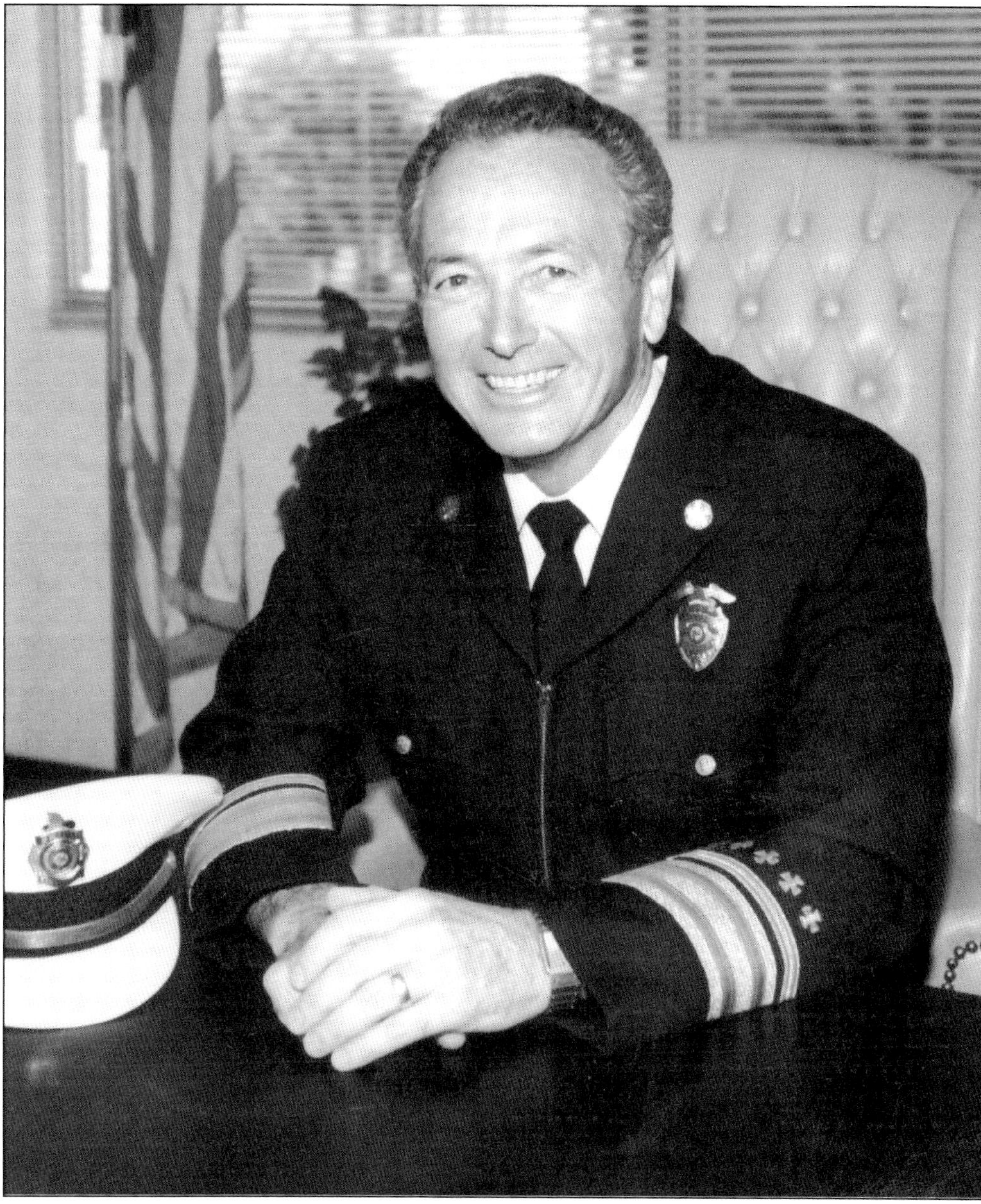

JIM REED: 1978 TO 1991. A veteran of the NBFD since 1961, Chief Jim Reed rose through the ranks to become chief in 1978. Chief Reed successfully guided the department through budgetary and labor ups and downs and laid the groundwork for future expansion into unincorporated areas destined for annexation. Administrative offices for the fire department were moved into city hall. A citywide emergency operational plan was adopted and an Emergency Operations Center was created at Fire Station Three; a Hazardous Materials Team was formed and trained; and mandatory EMT instruction and annual physical fitness testing for firefighters were instituted. During the 1980s, an Insurance Services Office evaluation of the Newport Beach Fire Department resulted in a class two rating, an exceptional grade and an achievement that was the source of much civic pride. (Courtesy of NBFD archives.)

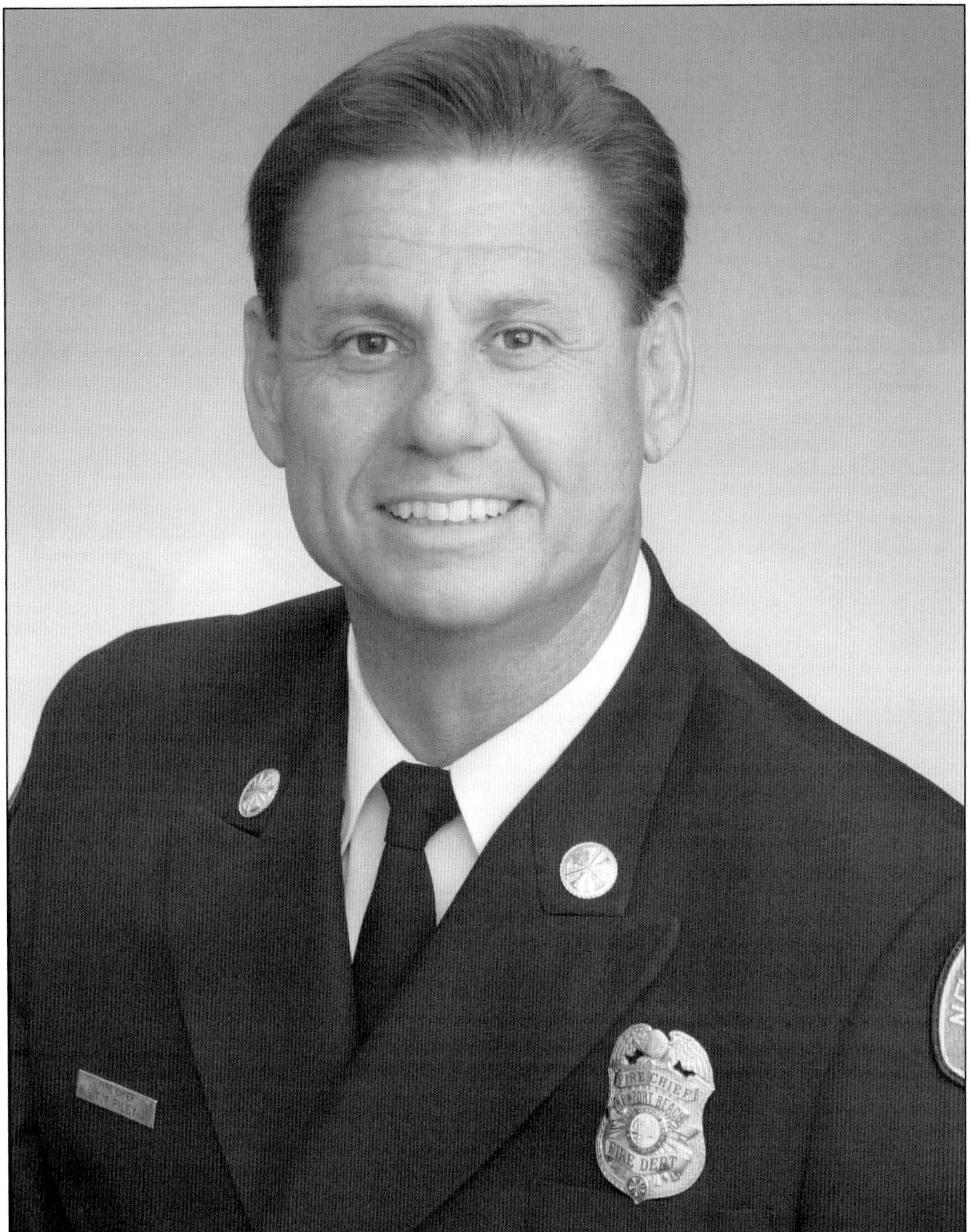

Tim Riley: 1991 to 2006. Tim Riley was recruited from the nearby Anaheim Fire Department. As the first fire chief not to come up through the NBFD ranks, Chief Riley brought a fresh perspective to fire department operations in Newport Beach. During Chief Riley's 15 years with Newport Fire, two new fire stations were added, lifeguards were brought back under fire department control, and fire dispatching was transferred to a state-of-the-art communications facility. A paramedic transport program was inaugurated and the fire apparatus fleet was uniformly upgraded. Disaster preparedness and fire/paramedic subscription service programs were implemented. (Courtesy of fire chief Timothy Riley.)

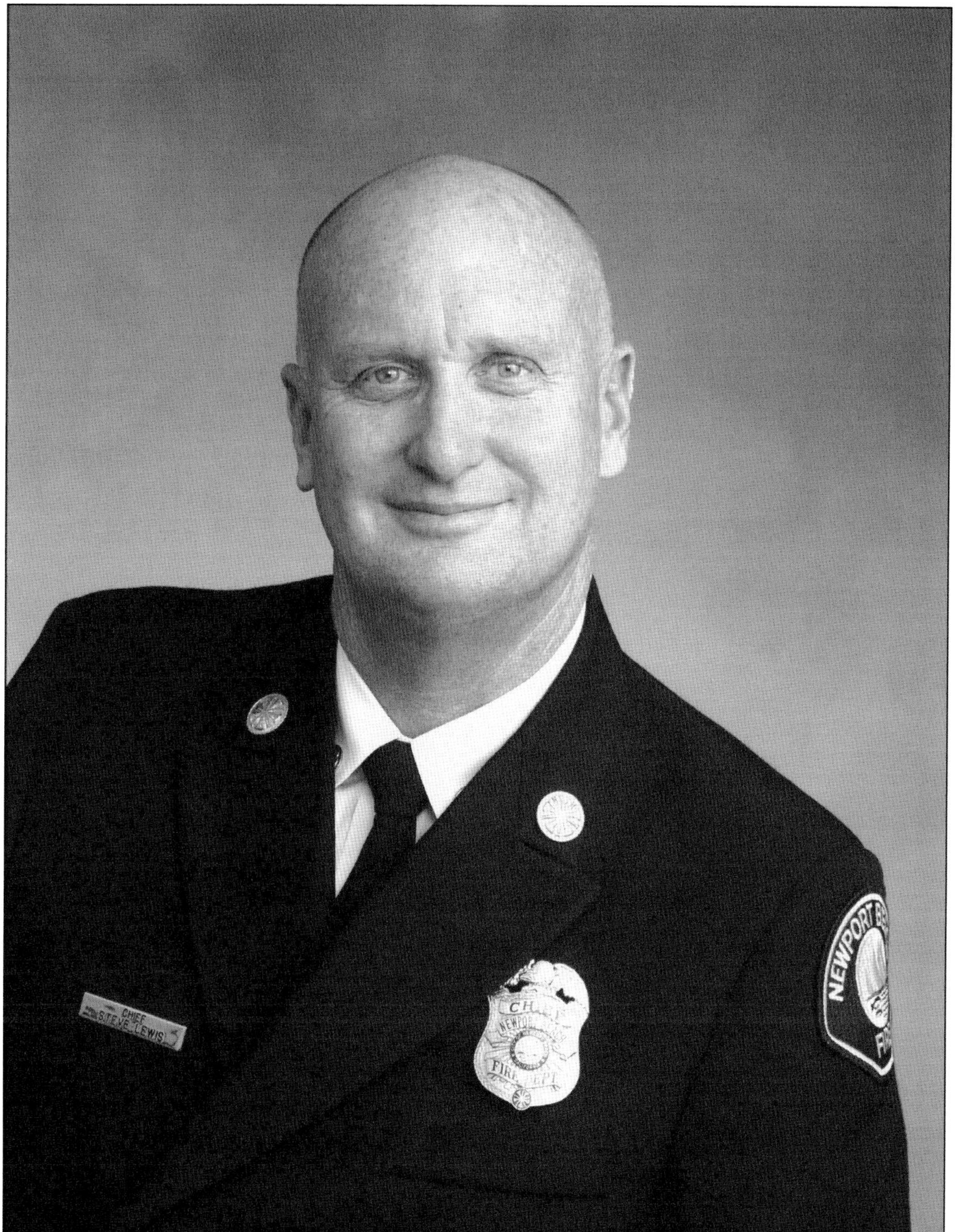

STEVE LEWIS: 2006 TO PRESENT. Fire chief Steve Lewis joined the Newport Beach Fire Department in 2006. Chief Lewis came to Newport Beach from the Long Beach Fire Department, where he was deputy chief. Way back in 1916, the Long Beach Fire Department provided Newport Fire with its first motorized fire apparatus and trained its first rescue-squad members. The Long Beach/Newport Fire connection has come full circle as Chief Lewis takes over the helm nearly 100 years later. (Courtesy of fire chief Steve Lewis.)